STP 1473

Beryllium: Sampling and Analysis

Dr. Kevin Ashley, editor

ASTM Stock Number: STP1473

ASTM International
100 Barr Harbor Drive
PO Box C700
West Conshohocken, PA 19428-2959

Printed in the U.S.A.

Library of Congress Cataloging-in-Publication Data

Symposium on Beryllium Sampling and Analysis (2005 : Reno, Nev.)
Beryllium : sampling and analysis / Kevin Ashley.
ISBN-13: 978-0-8031-3499-7 ISBN-10: 0-8031-3499-1
p. ; cm. — (STP ; 1473)
"Contains papers presented at the Symposium on Beryllium Sampling and Analysis, which was held in Reno, NV (USA) on 21-22 April, 2005.
The symposium was sponsored by ASTM International Committee D22 on Air Quality and its Subcommittee D22.04 on Sampling and Analysis of Workplace Atmospheres, in cooperation with the Sampling and Analysis Subcommittee of the Beryllium Health and Safety Committee"--Foreword.
1. Beryllium—Analysis—Congresses. I. Ashley, Kevin. II. ASTM International. Committee D22 on Air Quality. III. ASTM International.
Subcommittee D22.04 on Sampling and Analysis of Workplace Atmospheres. IV. ASTM International. Beryllium Health and Safety Committee. Sampling and Analysis Subcommittee. V. Title. VI. Series: ASTM special technical publication ; 1473.
[DNLM: 1. Beryllium--isolation & purification--Congresses. 2. Beryllium—analysis—Congresses. QV 275 S989b 2006]
QD181.B4S96 2006
615.9'25391—dc22 2006022213

Peer Review Policy

Each paper published in this volume was evaluated by two peer reviewers and at least one editor. The authors addressed all of the reviewers' comments to the satisfaction of both the technical editor(s) and the ASTM International Committee on Publications.

The quality of the papers in this publication reflects not only the obvious efforts of the authors and the technical editor(s), but also the work of the peer reviewers. In keeping with long-standing publication practices, ASTM International maintains the anonymity of the peer reviewers. The ASTM International Committee on Publications acknowledges with appreciation their dedication and contribution of time and effort on behalf of ASTM International.

Printed in Lancaster, PA
August, 2006

Foreword

This publication, *Beryllium: Sampling and Analysis*, contains papers presented at the Symposium on Beryllium Sampling and Analysis, which was held in Reno, NV (USA) on 21–22 April, 2005. The symposium was sponsored by ASTM International Committee D22 on Air Quality and its Subcommittee D22.04 on Sampling and Analysis of Workplace Atmospheres, in cooperation with the Sampling and Analysis Subcommittee of the Beryllium Health and Safety Committee. Dr. Kevin Ashley, Centers for Disease Control and Prevention / National Institute for Occupational Safety and Health, presided as symposium chairman and served as editor of this compilation. Co-chairs of the symposium were Kathryn L. Creek, Los Alamos National Laboratory; David Hamel, Occupational Safety and Health Administration; Michael J. Brisson, Washington Savannah River Company; and Dr. Amy Ekechukwu, Savannah River National Laboratory.

Kevin Ashley, Ph.D.
CDC/NIOSH, Cincinnati, OH
Symposium Chairman and Editor

Contents

ON-SITE MONITORING FOR BERYLLIUM – SAMPLING AND ANALYTICAL ASPECTS

Overview

This compilation represents the work of numerous authors at the Symposium on Beryllium Sampling and Analysis, April 21– 22, 2005, Reno, Nevada. The symposium was sponsored by ASTM International Committee D22 on Air Quality and its Subcommittee D22.04 on Workplace Atmospheres, in cooperation with the Sampling and Analysis Subcommittee of the Beryllium Health and Safety Committee. Eighteen papers were presented at the symposium, and nine presentations that were accepted for publication appear in this volume.

Occupational exposure to beryllium can cause a lung disease that may ultimately be fatal, and new exposure limits for this element in air and on surfaces have been established in efforts to reduce exposure risks to potentially affected workers. Advances in sampling and analytical methods for beryllium are needed in order to meet the challenges relating to exposure assessment and risk reduction. This symposium provided a forum for technical exchanges on current research and status regarding beryllium sampling and analysis issues. While the primary emphasis was on current research in the areas of beryllium sample collection, sample preparation, and measurement, participants were able to identify areas where pertinent standards can be developed concerning beryllium sampling and analytical procedures.

The symposium was intended to address topics related to: 1. Sampling of beryllium in workplace atmospheres; 2. Surface beryllium sampling; 3. Sample preparation procedures for beryllium in various matrices; 4. Analytical methods for measuring beryllium; 5. Beryllium reference materials; 6. beryllium proficiency testing; 7. On-site beryllium monitoring; and 8. Opportunities for standardization of beryllium sampling and analysis methods. The targeted audience included technical professionals such as industrial hygienists, chemists, health physicists, safety engineers, epidemiologists, and others having interest in beryllium exposure and analysis issues.

The papers contained in this publication represent the commitment of the ASTM D22.04 subcommittee to providing timely and comprehensive information on advances in workplace exposure monitoring. Sections of the two-day symposium focused on the following themes: 1. Beryllium disease – Exposure monitoring and standardization issues; 2. Beryllium exposure measurement and reference materials – National and international perspectives; and 3. On-site monitoring for beryllium – Sampling and analytical aspects. Papers discussing beryllium sampling techniques, analytical measurement technologies, beryllium reference materials, standardization, and occupational hygiene can be found in this compilation.

Beryllium disease – Exposure monitoring and standardization issues

The intent of this section was to present an overview of beryllium disease and efforts to reduce worker exposures through improved monitoring methods and the development of standard methodologies. Some of the papers presented discussed the industrial uses of beryllium and the history of beryllium disease. Other papers dealt with occupational exposure monitoring and standardization of sampling and analytical methods. These areas continue to comprise the activities of many beryllium researchers. Two of the presented papers from this section of the symposium are published herein.

Beryllium exposure measurement and reference materials – National and international perspectives

This portion of the symposium covered global efforts and progress in beryllium occupational monitoring, as well as the development and characterization of beryllium reference materials. Applications of sampling and analytical methods to industrial hygiene chemistry and practice were highlighted, and needs for reference materials containing beryllium oxide were identified. Four of the papers that were given dealing with these issues are published in this section.

On-site monitoring for beryllium – Sampling and analytical aspects

The ability to carry out on-site beryllium analysis has been a desire of many for years, and this part of the symposium covered recent developments in this area. New portable analytical methods for determining trace beryllium in samples from air and surfaces have been developed and evaluated, and advances in this research arena are continuing. These include both real-time qualitative and semi-quantitative methods, as well as near real-time quantitative techniques for ultra-trace beryllium analysis. Three papers that were presented in this part of the symposium are published here.

Kevin Ashley
CDC/NIOSH, Cincinnati, OH
Symposium Chairman and Editor

Acknowledgments

The editor gratefully acknowledges the voluntary contributions of the numerous colleagues who served as peer reviewers of the manuscripts that were submitted for consideration for publication. Their efforts made the symposium and this compilation possible. Special thanks are extended to the following symposium co-chairs, who helped arrange the presentations and kindly served as session monitors:

Kathryn L. Creek
Los Alamos National Laboratory
Los Alamos, NM

David Hamel
Occupational Safety and Health Administration
Washington, DC

Michael J. Brisson
Washington Savannah River Company
Savannah River Site, SC

Amy Ekechukwu
Savannah River National Laboratory
Savannah River Site, SC

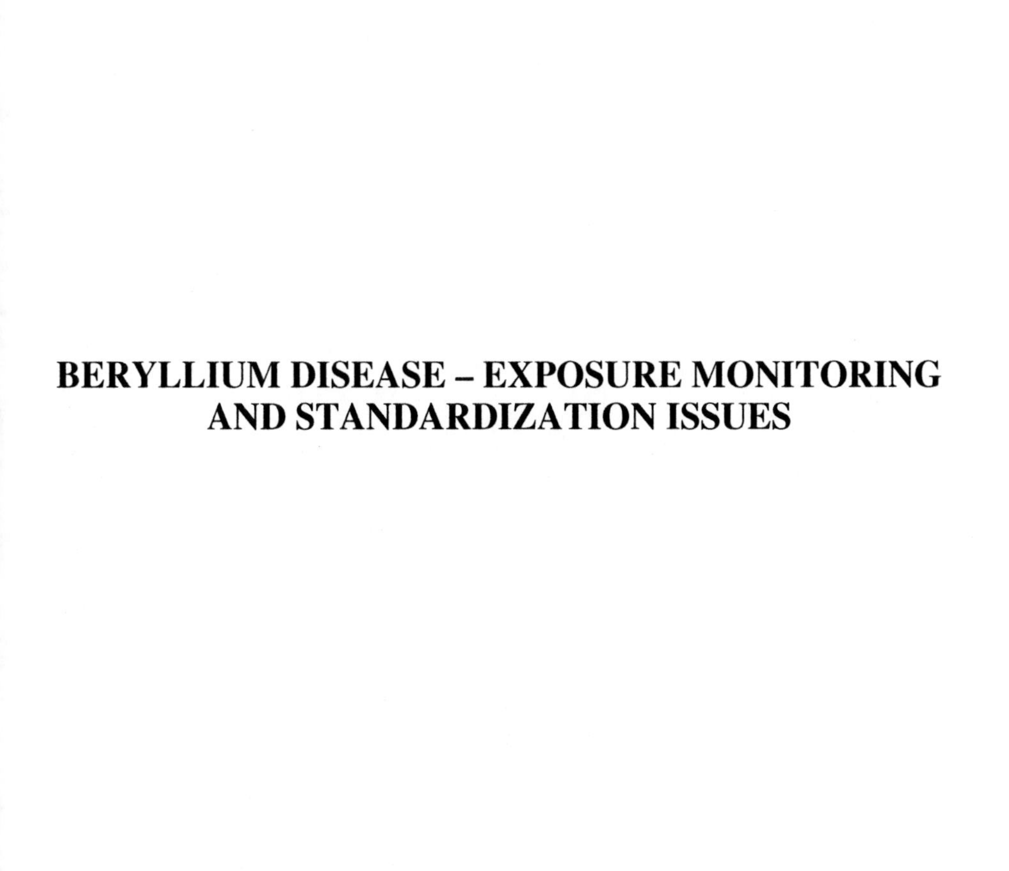

BERYLLIUM DISEASE – EXPOSURE MONITORING AND STANDARDIZATION ISSUES

Journal of ASTM International, January 2006, Vol. 3, No. 1
Paper ID JAI13157
Available online at www.astm.org

Michael J. Brisson, M.Sc.,[1] *Amy A. Ekechukwu, Ph.D.,*[2] *Kevin Ashley, Ph.D.,*[3] *and Steven D. Jahn, CIH*[4]

Opportunities for Standardization of Beryllium Sampling and Analysis*

ABSTRACT: Since the U. S. Department of Energy (DOE) published the DOE Beryllium Rule (10 CFR 850) in 1999, DOE sites have been required to measure beryllium in air filter and surface wipe samples for purposes of worker protection and for release of materials from beryllium-controlled areas. Measurements in the nanogram range on a filter or wipe are typically required. Industrial hygiene laboratories have applied methods from various analytical compendia, and a number of issues have emerged concerning sampling and analysis practices. As a result, a committee of analytical chemists, industrial hygienists, and laboratory managers was formed in November 2003 to address the issues. The committee developed a baseline questionnaire and distributed it to DOE sites and other agencies in the U.S., Canada, and the U.K. The results of the questionnaire are presented in this paper. These results confirmed that a wide variety of practices was in use in the areas of sampling, sample preparation, and analysis. Additionally, although these laboratories are generally accredited by the American Industrial Hygiene Association (AIHA), there are inconsistencies in execution among accredited laboratories. As a result, there are significant opportunities for development of standard methods that could improve consistency. The current availabilities and needs for standard methods are further discussed in a companion paper.

KEYWORDS: analysis, beryllium, sampling, standards, workplace

Introduction

Beryllium metal, oxide, and alloys have been used for many years in such diverse applications as aerospace, nuclear weapons, automotive, and sports equipment [1]. Unfortunately, exposure to these forms of beryllium through inhalation [2] or dermal exposure [3] can lead to sensitization and, in a small percentage of those sensitized, to chronic beryllium disease (CBD). Symptoms of CBD may not appear until 10–15 years after exposure. CBD can result in sarcoidosis (granulomatous lesions) in the lungs and is treatable but not curable [1]. For these reasons, workplace monitoring is required where beryllium particles can become airborne or deposited on accessible surfaces.

The U.S. Department of Energy (DOE), in response to growing concerns about workplace exposure to beryllium in its nuclear weapons facilities, published its Chronic Beryllium Disease Prevention Program (CBDPP), 10 CFR 850 [4] in December 1999 (also known as the DOE

Manuscript received 23 February 2005; accepted for publication 27 June 2005; published January 2006. Presented at ASTM Symposium on Beryllium: Sampling and Analysis on 21-22 April 2005 in Reno, NV.
[1] Westinghouse Savannah River Company, Savannah River Site, Aiken, SC 29808 (USA). Author for correspondence: tel. +1(803)952-4402; fax +1(803)952-3063; e-mail: mike.brisson@srs.gov.
[2] Savannah River National Laboratory, Aiken, SC 29808 (USA).
[3] U.S. Department of Health and Human Services, Centers for Disease Control and Prevention, National Institute for Occupational Safety and Health, Cincinnati, OH 45226-1998 (USA).
[4] Westinghouse Savannah River Company, Savannah River Site, Aiken, SC 29808.

Beryllium Rule). As part of the CBDPP established under the Rule, monitoring requirements were imposed that included sampling for beryllium in workplace atmospheres and surfaces, both for purposes of worker protection and for release of materials from beryllium-handling areas. The resulting air filter and surface wipe samples are analyzed in a laboratory accredited for metals analysis by the American Industrial Hygiene Association (AIHA) or by a laboratory that can demonstrate a quality assurance program equivalent to that required by AIHA accreditation. The intent of this requirement was to ensure the quality of the analytical results, and to allow for comparison of results from site to site. Because the action level is 0.2 μg per m^3 of air or per 100 cm^2 of surface, laboratories typically need to be able to measure beryllium in the nanogram range on air filters or surface wipes.

In 2002, issues (as discussed below) began to be identified with the analyses being performed by laboratories supporting DOE sites. It was discovered that, although AIHA-accredited laboratories were being used, there was a wide variety of sample collection, sample preparation, and analysis protocols being employed. As a result, a questionnaire was developed and distributed to a number of sites, including DOE sites and other facilities performing similar work in the U.S. and the U.K., to compile information on the protocols being used. This paper presents the information obtained from that questionnaire. The results demonstrate an opportunity for further method development and standardization and for development of additional standard reference materials. It is believed that better harmonization of laboratory protocols could improve the consistency of sampling and analytical results from different sites.

Questionnaire Background and Development

Differences among the responding sites focus on three major areas: sampling, sample preparation, and analysis. In air sampling, the principal difference is in the air filter media being employed. In wipe sampling, some sites use wetted wipes, while others use dry wipes, with resulting variations in collection efficiency [5]. Additional sampling issues, such as air volumes collected, statistical sampling plans, and bulk sampling, are outside the scope of this paper.

Sample preparation requires digestion of the filter or wipe media, typically in an acid matrix at high temperature and/or pressure. Laboratories typically use a digestion protocol based on a published standard method (e.g., ASTM, U.S. National Institute for Occupational Safety and Health [NIOSH], U.S. Environmental Protection Agency [EPA], U.S. Occupational Safety and Health Administration [OSHA], International Organization for Standardization [ISO], U.K. Health and Safety Executive [HSE]), with modifications in some cases. These standard protocols typically will digest many of the forms of beryllium encountered at worksites, and will also digest beryllium acetate, which is the form of beryllium currently used in AIHA Beryllium Proficiency Analytical Testing Program (BePAT) samples.[5] However, results vary among these methods with respect to digestion of more refractory forms of beryllium, such as beryllium oxide (BeO). Some methods, such as NIOSH Method 7300 [6], provide disclaimers about their ability to digest certain compounds of beryllium. As a result, there may be uncertainty about whether full recovery of all beryllium species is being achieved. Studies are limited due largely to the lack of a BeO standard reference material [7].

Analysis methods are typically based on spectrometric techniques such as inductively coupled plasma atomic emission spectrometry (ICP-AES) or graphite furnace atomic absorption spectrometry (GFAAS). Inductively coupled plasma mass spectrometry (ICP-MS) is not widely

---

[5] Personal communication, M. J. Brisson to L. D. Welch (BWXT Y-12), February 2, 2005.

used, but because it offers a detection limit roughly one order of magnitude lower than ICP-AES or GFAAS [8,9], some sites use ICP-MS when a lower detection limit is required. In ICP-AES, spectral interferences may be encountered, which, if not properly corrected, can cause inaccurate results. In 2002, a number of samples from Savannah River Site (SRS) were sent to two AIHA-accredited laboratories, which reported different results due to disparities in both interference correction protocols and sample preparation (digestion) protocols [10].

The above differences came to light in discussions among industrial hygiene and analytical laboratory personnel at various DOE sites. The principal forum for sharing information has been an ad hoc group known as the Beryllium Health and Safety Committee (BHSC), which includes representatives from DOE sites, NIOSH, OSHA, the U.S. Department of Defense (DOD), and the U.K. Atomic Weapons Establishment (AWE). In November 2003, the BHSC formed an Analytical Subcommittee to improve the consistency and quality of sampling and analysis methodologies and to enhance communication between industrial hygiene and analytical laboratory personnel at participating sites [24]. In March 2004, the Subcommittee issued a questionnaire to BHSC member sites to collect information regarding sampling, sample preparation, and analysis protocols being used. A total of 16 responses was received (14 from U.S. locations and one each from the U.K. and Canada).

Questionnaire Results

Table 1 provides background information from the responding sites. This includes information about sample volumes, whether any samples are radiologically contaminated, whether processes involving BeO are used, and accreditation status of analytical laboratories. Sites are identified by country or by U.S. agency, but they are not identified by name.

The results indicate a wide variety of sample volumes. Eight of the 16 respondents have radiologically-contaminated beryllium samples. Three have processes involving BeO; in these cases, the firing temperatures used were either not available or not provided for security reasons. All non-radiological laboratories are accredited by AIHA or HSE. Three radiological laboratories are AIHA accredited; one has an equivalent quality assurance program, and four are not accredited.

Sampling Protocols

With respect to sampling protocols, the focus of the questionnaire was on surface wipe sampling; for that reason, only six of the responding sites provided information on air sampling. Of the six laboratories reporting on air sampling media, five of them use 0.8 mm mixed cellulose ester (MCE) filters, while one site uses Whatman® 41 ashless cellulose filters.

Information provided for surface wipe samples is provided in Table 2. A wide variety of collection media is employed; several sites use multiple media types. The media type most frequently used is Ghost Wipes® (Environmental Express) [11], which is compliant with ASTM Standard Specification for Wipe Sampling Materials for Lead in Surface Dust (E 1792). ASTM E 1792 is referenced in ASTM Standard Practice for Collection of Settled Dust Samples Using Wipe Sampling Methods for Subsequent Determination of Metals (D 6966). The questionnaire asks for information about pore size; however, this information is not available for Ghost Wipe® media[6], nor is pore size specified by ASTM E 1792 or ASTM D 6966. Whatman® filters are the next most frequently used media (primarily Whatman® 41).

---

[6] Personal communication, M. J. Brisson to Robert Benz (Environmental Express), January 24, 2005.

The use of wet or dry collection methods has been a major source of discussion among DOE sites performing beryllium analyses. It is noted in Table 2 that, of the 16 respondents, ten use only wetted wipes (water, alcohol, or other organic agent), two use only dry wipes, and four use both types depending on the specific application. Use of dry wipes is typically based on historic practices, which at DOE sites are often based on wipes used for radioactive surface contamination (which are dry). Dry wipes are also required in some cases to avoid damage to the surface being wiped. Advocates of wet wipes typically cite better collection efficiencies; however, as can be seen by Table 2, few collection efficiency studies have been performed.

Collection methods also vary widely and include NIOSH method 9100 [12], ASTM D 6966, ASTM Standard Practice for Collection of Settled Dust Samples Using Wipe Sampling Methods for Subsequent Lead Determination (E 1728), guidelines published in 1995 from the U.S. Department of Housing and Urban Development (HUD) [13], OSHA [14], and unpublished in-house methods. As noted in Table 2, a combination of methods and/or modifications to published methods is used at some sites. Even when the same collection method is used, human variability can have an impact on variability of results. The variety of collection methods further increases variability. It is pointed out that some of these methods were developed for lead sampling and are now being applied to beryllium sampling. In most cases, data have not been collected to demonstrate that these surface sampling methods provide performance for beryllium that is comparable to the performance measured for lead. It should be noted that, subsequent to the questionnaire responses, NIOSH has published Method 9102 [15], which updates NIOSH Method 9100 to include beryllium and other elements.

TABLE 1—*Background information by site.*

Site ID[a]	# Air Samples per year	% Rad Air Samples	# Wipe Samples per year	% Rad Wipe Samples	Hot Processes (>500°C)	# BeO Proc.	Accreditation[b]
Can-1	650	0	2500	0	0	0	AIHA
DOD-1	44	0	0	0	0	0	AIHA
DOD-2	775	0	3	0	0	0	AIHA
DOD-3	150	0	30	0	0	0	AIHA
DOE-1	200	...	1400	...	0	...	AIHA (both)
DOE-2	37	0	184	11	0	...	AIHA (non-rad)
DOE-3	2522	<1	7746	24	6	0	AIHA (both)
DOE-4	2200	...	13000	14	3	1	AIHA (non-rad)
DOE-5	243	43	329	...	Yes	...	AIHA (non-rad)
DOE-6	269	25	20500	1	0	...	AIHA (non-rad)
DOE-7	50	0	500	0	3	1	AIHA
DOE-8	50	13	600	7	0	0	AIHA (non-rad); equivalent (rad)
DOE-9	6175	13	33250	18	0	0	AIHA (both)
NIOSH-1	...	0	...	0	0	0	AIHA
OSHA-1	...	...	4280[c]	0	3	1	AIHA
UK-1	12000	33	17000	43	3	0	HSE

[a] Sites are identified as to whether they are Canadian (Can), U.S. National Institute for Occupational Safety and Health (NIOSH), Department of Defense (DOD), U.S. Department of Energy (DOE), U.S. Occupational Safety and Health Administration (OSHA), or U.K.

[b] "Rad" refers to radioactive laboratories, while "non-rad" refers to non-radioactive laboratories. Sites with radiologically-contaminated samples typically analyze them in a different location from the non-rad samples. If "non-rad" is denoted, only that laboratory is accredited; "both" means that both rad and non-rad laboratories are accredited.

[c] Only a combined value for air and wipe samples was provided.

TABLE 2—*Surface wipe characteristics by site.*

Site	Media Type	Dry or Wet & Wetting Agent	Collection Efficiency Study?	Collection Method	Reference Materials Used
Can-1	Ghost Wipe®	Water	No	NIOSH 9100/ ASTM D 6966	Spex® standard solutions
DOD-1	Ghost Wipe®	Wet (agent not named)	N/A	...	...
DOD-2	Ghost Wipe®	Alcohol	No	HUD (1995)/ OSHA 125G	AIHA Proficiency Samples
DOD-3	Ghost Wipe®	Alcohol	No	HUD (1995)/ OSHA 125G	AIHA Proficiency Samples
DOE-1	Ghost Wipe®	Organic	No	ASTM D 6966	None routinely
DOE-2	6x6 Gauze	Methanol	No	EPA 6010	None routinely
DOE-3	Whatman® 541 or 41, or linen cloth	Both dry and wet (water)	In progress	NIOSH 9100 (modified)	BeO performance samples
DOE-4	Whatman® 50 smear tab	Dry	No	In-House	...
DOE-5	Smear tab	Water	...	ASTM E 1728	...
DOE-6	Whatman® 41 filter	Both dry and wet (water)	No	NIOSH 9100 (modified)	N/A
DOE-7	Ghost Wipe®, Whatman® 41, or smear tab	Both dry and wet (water)	No	NIOSH 9100	None
DOE-8	Ghost Wipe®	Water	No	...	High Purity Filters
DOE-9	Whatman® 41, Ghost Wipe®	Both dry and wet (water)	Yes	...	AIHA PAT
NIOSH-1	Ghost Wipe®	Water	...	NIOSH 9100	Analytical standards, spiked wipes/filters, BeO suspensions
OSHA-1	Smear Tabs, Whatman® 41 and 42, Ghost Wipe®	Water	Yes	OSHA 125G	None
UK-1	Whatman® 41	Dry	Yes	In-House	...

Similarly, a variety of responses was given to the question, "Which reference sample materials are employed?," as shown in the last column of Table 2. Respondents were asked what reference materials are lacking and need to be produced. Most sites indicated a need for proficiency test samples containing BeO; however, the lack of a BeO reference material makes it impossible to develop such samples at present.

It should be noted that even when variables such as media type and collection method are eliminated, typical sampling uncertainty is greater than analytical uncertainty. The large variety of media types and collection methods, the lack of collection efficiency data, and differences in reference materials all make it difficult to compare sampling results. These issues present opportunities for standardization that are discussed in a companion paper [16].

Sample Preparation Protocols

Analytical techniques that meet the performance requirements of the DOE Beryllium Rule [4], or similar performance requirements such as the ability to measure at or below the Threshold Limit Value (TLV) published by the American Conference of Governmental Industrial Hygienists (ACGIH) [17], require that the sample be dissolved prior to analysis. To date, no direct-solid measurement technique has been validated to meet these requirements [18]. Therefore, sample preparation is necessary prior to analysis for beryllium.

Table 3 displays information on sample preparation used at the responding sites for air filter samples. Table 4 displays similar information for surface wipe samples. Differences between the two tables are highlighted in Table 4.

TABLE 3—*Sample preparation techniques for air filter samples, by site.*

Site ID[a]	Energy System	Digestion Reagents	Final Sample Volume (mL)	Storage time (typical/maximum)
Can-1	Hotplate	HNO_3, $HClO_4$	10	2-3 days/ ...
DOD-1	Hot block	HNO_3, H_2O_2	50	1-2 weeks/ ...
DOD-2	Open vessel (OV) microwave	HNO_3, H_2O_2, HCl	25	4 hours/1 week
DOD-3	Hot block or OV microwave	HNO_3, H_2O_2	15-25	4-16 hours/3 days
DOE-1	Hotplate	HNO_3, HCl	10	1-7 days/2 weeks
DOE-2	Closed vessel (CV) microwave	HNO_3	...	<1 day/ ...
DOE-3	Hotplate	H_2SO_4, HNO_3, H_2O_2, HCl	10	<1 day/2 weeks
DOE-4	OV microwave	H_2SO_4	25	1-2 days/ ...
DOE-6	Hot block	HNO_3, H_2SO_4, $HClO_4$	25	1 hour/<1 day
DOE-7	CV microwave	HNO_3	25	24 hours/30 days
DOE-8	Hot block	HNO_3, H_2O_2, HCl, HF	25	24 hours/> 2 weeks
DOE-9	OV microwave	H_2SO_4, HNO_3	10	<1 day/14 days
NIOSH-1	Hotplate	HNO_3, $HClO_4$	10	1 day/ ...
OSHA-1	Hotplate	H_2SO_4, HNO_3, H_2O_2, HCl	50	1 day/15 days
UK-1	Hotplate	HNO_3, $HClO_4$	5	<1 week/<2 weeks

[a]Site DOE-5 did not respond to this portion of the questionnaire.

TABLE 4—*Sample preparation techniques for surface wipe samples, by site.*

Site ID[a]	Energy System	Digestion Reagents	Final Sample Volume (mL)	Storage time (typical/maximum)
Can-1	Hotplate	HNO_3, H_2O_2 [b]	10	2-3 days/ ...
DOD-1	OV microwave [b]	HNO_3 [b]	100 [b]	1-2 weeks/ ...
DOD-2	OV microwave	HNO_3, H_2O_2 [b]	100 [b]	1 week [b] /1 week
DOD-3	Hot block	HNO_3, H_2O_2	50 [b]	1-2 days [b]/1 week [b]
DOE-1	Hotplate	HNO_3, HCl	10	1-14 days [b] /28 days [b]
DOE-2	Hot block [b]	HNO_3, HCl [b]	100	<1 day/ ...
DOE-3	Hotplate	H_2SO_4, HNO_3, H_2O_2, HCl	10	<1 day/2 weeks
DOE-4	OV microwave	H_2SO_4, H_2O_2 [b]	25	1-2 days/ ...
DOE-6	Hot block	HNO_3, H_2SO_4, $HClO_4$	50 [b]	1 hour/<1 day
DOE-7	OV microwave [b]	HNO_3, H_2O_2 [b]	50 [b]	24 hours/30 days
DOE-8	Hot block	HNO_3, H_2O_2, HCl, HF	25	<24 hours/48 hours
DOE-9	OV microwave	H_2SO_4, HNO_3, H_2O_2 [b]	10	<3 days [b] /10 days [b]
NIOSH-1	Hotplate	HNO_3, $HClO_4$	10	1 day/ ...
OSHA-1	Hotplate	H_2SO_4, HNO_3, H_2O_2, HCl	50	1 day/ ...
UK-1	Hotplate	HNO_3, $HClO_4$	5	<1 week/ ...

[a]Site DOE-5 did not respond to this portion of the questionnaire.
[b]Response differs from that given in Table 3.

As noted in both tables, energy systems include hotplate, hot block, and microwave (open or closed vessel). Some sites use more than one system for air filter samples. Three sites use a different system for surface wipe samples from that used for air filter samples.

Reagent protocols vary widely, and half of the respondents use a different protocol for surface wipe samples from that used for air filter samples. Typically, where this is the case, the surface wipe protocol features a more robust acid combination than that used for air filters, since the latter are typically easier to digest. In addition, a range of heating energy systems and dissolution times occurs across the respondents.

Additional variation exists in how much of each acid is used in each site's sample preparation. This information was captured in the questionnaire responses but, for simplicity, is not reported in Tables 3 and 4.

Final sample volumes also vary widely; five of the respondents use greater sample volumes for surface wipe samples than for air filter samples. This again is a function of the greater difficulty in digesting wipe media. A review of Table 4 indicates that when considering energy system, reagents, and final sample volume, each of the responses is unique. This is close to being true for Table 3 as well (note that Can-1 and CDC-1 are the same for these three parameters, but these two sites vary in their acid concentrations). Although there may not necessarily be a "one-size-fits-all" sample preparation approach that would meet everyone's needs, this still appears to present a fertile opportunity to improve consistency.

Storage time is simply an indication of the length of time, both typically and in the maximum case, between sample preparation and analysis, based on each lab's actual experience. There are no official "hold times" such as that typically found for environmental samples, and to our knowledge there has not been any detailed study to support any particular duration. The experience of the respondents suggests that prepared samples can be held up to 30 days before being analyzed, but that two weeks or less is more typical.

Analysis Protocols

Table 5 displays information on analysis methods used at the responding sites for air filter samples. Table 6 displays similar information for surface wipe samples. Tables 5 and 6 will be discussed together.

Both Table 5 and Table 6 indicate that ICP-AES is used by the majority of responding sites (11 of 15 for air filters, 12 of 15 for surface wipes), either exclusively or as the primary instrument. ICP-MS is used as the primary instrument at a few sites (two for air filters, three for surface wipes) and as a backup instrument at one site. One respondent uses GFAAS.

Tables 5 and 6 indicate that a wide variety of methods is in use. While NIOSH method 7300 [6] is widely used (eight respondents), it is modified in some way at several sites (two for air filters, five for surface wipes). Other methods used include EPA methods 200.8 [9], 6010B [19], and 6020 [20]; OSHA method ID-125G [14]; and in-house methods. It should be noted that all of the standard methods listed are for various suites of elements, not specifically for beryllium, and they may not necessarily be optimized for beryllium at trace levels. It is notable that NIOSH Method 7102 [21], which is specific for trace-level beryllium, is not cited. This is probably due to the fact that NIOSH Method 7102 is a GFAAS method, and GFAAS is not used by any of the U.S. respondents.

The questionnaire asked for detection limits for both air filters and surface wipes, and how the detection limits were determined. The responses varied widely both in terms of numerical values and units of measure. Although some of the variation can be attributed to differences in sample preparation (see Tables 3 and 4), sample matrices, and analytical instrumentation, there are also differences in how detection limits are calculated. Various organizations (e.g., NIOSH, EPA, ASTM, ISO, American Chemical Society) have promulgated different methodologies for computing detection limits; a variety of these are used by the respondents. Also, the questionnaire did not distinguish between instrument detection limit (IDL) and method detection limit (MDL), so the values provided are likely a mix of both types.

TABLE 5—*Analytical methods for air filter samples, by site.*

Site ID[a]	Instrument	Method	Detection Limit	Reporting Limit	Line(s) Used (nm)	Internal Standard	Known Interferences
Can-1	ICP-MS	In-House	0.0005 µg/sample	0.0005 µg/sample	N/A	Lithium, Scandium	...
DOD-1	ICP-MS	NIOSH 7300 (mod.)	0.0085 µg/filter	0.25 µg/filter	N/A	Lithium	...
DOD-2	ICP-AES	NIOSH 7300	0.05 µg/sample	...	313.107	...	...
DOD-3	ICP-AES	NIOSH 7300	0.01 µg/sample	0.02 µg/sample	...	None	None
DOE-1	ICP-MS	EPA 200.8 (mod.)	0.0007 µg/filter	0.005 µg/filter	N/A	Scandium	None
DOE-2	ICP-AES	NIOSH 7300	...	0.02 µg/filter	...	...	V
DOE-3	ICP-AES	OSHA ID-125G (mod.)	0.003 µg/filter	0.03 µg/filter	313.107	Yttrium	Al, Cu, Fe, V
DOE-4	ICP-AES	In-House	0.02 µg	...	Three different lines (not identified)	Yttrium	...
DOE-6	ICP-AES	NIOSH 7300 (mod.)	0.12 ppb	0.01 µg/filter	234.861	None	Fe
DOE-7	ICP-AES/ ICP-MS	NIOSH 7300/ EPA 6020	0.003 µg/sample (AES); 0.00063 µg/sample (MS)	...	313.042 (AES)	Scandium or Lithium (MS only)	V (AES only)
DOE-8	ICP-AES	NIOSH 7300 mod/ EPA 6010B	0.005 µg/sample	0.05 µg/sample	313.107	...	Fe, Ti
DOE-9	ICP-AES	In-House	0.144 µg/L	1.0 µg/L	313.042 313.107	Scandium	...
NIOSH-1	ICP-AES	NIOSH 7300	0.005 µg/sample	...	313	None	...
OSHA-1	ICP-AES	OSHA ID-125G	0.017 µg	0.02 µg/mL	313.107	...	Fe, Mn, Mo, Nb, Ni, Ti, V
UK-1	GFAAS	...	2.5 ng/sample	...	234.9	None	...

[a]Site DOE-5 did not respond to this portion of the questionnaire.

TABLE 6—*Analytical methods for surface wipe samples, by site.*

Site ID[a]	Instrument	Method	Detection Limit	Reporting Criteria	Line(s) Used (nm)	Internal Standard	Interferences Tested
Can-1	...	...	0.005 µg/sample	...	...	Lithium, scandium	...
DOD-1	ICP-MS	EPA 6020	0.19 µg/wipe	CFR	N/A	Lithium	...
DOD-2	ICP-AES	NIOSH 7300	0.1 µg	LOD	313.107	...	None routinely
DOD-3	ICP-AES	NIOSH 7300	0.1 µg	LOD	...	...	None routinely
DOE-1	ICP-AES	NIOSH 7300	0.02 µg/wipe	LOQ	313.042, 234.861	None	Elements not specified
DOE-2	ICP-AES	EPA 6010B	0.1 µg (RL)	LOQ	313.042	None	Al, Fe, Mn, Ca, Mg, Cd, Cu, Cr, Pb, Ti
DOE-3	ICP-AES	NIOSH 7300	0.01 µg/sample	... (0.02 µg/sample)	313.107	Yttrium	Al, Cu, Fe, V
DOE-4	ICP-AES	In-House	0.02 µg	...	...	Scandium	None
DOE-6	ICP-AES	NIOSH 7300 (Mod.)	0.12 ppb	LOQ (0.03 µg/wipe)	234.861	None	Fe, Mg, Ca
DOE-7	ICP-AES/ ICP-MS	NIOSH 7300/EPA 6020	0.0061 µg/sample (AES); 0.0013 µg/sample (MS)	LOQ	313.042 (AES)	Scandium or Lithium (MS only)	V (AES only)
DOE-8	ICP-AES	NIOSH 7300 mod/ EPA 6010B	0.005 µg/sample	PQL (0.05 µg/sample)	313.107	...	Al, As, B, Ba, Ca, Cd, Ce, Co, Cr, Cu, Fe, K, Mg, Mn, Na, Ni, P, Pb, Se, Sr, Tl, V, Zn
DOE-9	ICP-AES	In-House	0.144 µg/L	...	313.042 313.107	Scandium	Cr, Fe, Mo, Th, Ti, U, V, Y, Zr
NIOSH-1	ICP-AES	NIOSH 7300	0.005 µg/sample	LOQ, LOD	313	None	None
OSHA-1	ICP-AES	OSHA ID-125G	0.017 µg	...	313.107 234.861	...	Al, Ce, Cr, Co, Cu, Fe, Mn, Mo, Nb, Ni, Pt, Sb, V, Ti
UK-1	GFAAS	...	0.2 µg/sample	...	234.9	None	...

[a]Site DOE-5 did not respond to this portion of the questionnaire.

For air filter samples, the questionnaire also requested reporting limits (RL), which are shown in Table 5. Again, because of differences in how RLs are calculated by each lab, the values vary widely. Terminology is also an issue, since the organizations cited above use different terms. This is further illustrated in Table 6. For surface wipe samples, the questionnaire asked for reporting criteria rather than RLs. A variety of terms are used, including Limit of Quantitation (LOQ), Limit of Detection (LOD), and Practical Quantitation Limit (PQL), definitions of which can be found in a recent EPA comparison of detection and quantitation approaches [22].[7]

A significant issue with ICP-AES is spectral interferences. Most spectral lines have one or more spectral interferences that, if present in the sample, may affect the results. Proper correction for such interferences is important. For air filters, we asked for information on known interferences, which are presented in Table 5. In air samples, the list of known interferences appears to be relatively short. For surface wipe samples, we asked for information on the interferences that are tested. Based on the results (shown in Table 6), interference testing is a greater concern for surface wipe samples than for air filters.

Tables 5 and 6 also list the spectral line(s) used by each respondent for ICP-AES or GFAAS and whether internal standards are applied.

Path Forward

The need for standardization of sampling, sample preparation, and analysis techniques is clearly apparent. This topic is further discussed in a companion paper [16]. The BHSC Analytical Subcommittee is currently working through the ASTM Subcommittee on Workplace Atmospheres (D22.04) to develop some of these standards (http://www.astm.org/). The BHSC Analytical Subcommittee is also working toward greater consistency in how detection and reporting limits are computed. The Subcommittee and its members function as an ad hoc group and do not perform any advisory functions for DOE or other government agencies.

DOE is currently drafting a Technical Standard that is intended to recommend selected sampling and analytical methods to DOE laboratories as a means of improving consistency within the DOE complex. The DOE Technical Standards Program promotes the use of voluntary consensus standards at DOE [23].

A number of potential research opportunities is also being pursued. Among these are development of a BeO reference material (which should lead to improved proficiency test samples), improved sample preparation techniques, and studies on existing sampling techniques including collection efficiencies. Efforts are also being made toward field-deployable beryllium analyzers [18].

The above efforts are needed to improve sampling and analytical methodologies and allow for better comparisons of data among laboratories performing beryllium analyses for worker protection purposes.

Acknowledgments

The authors gratefully acknowledge members of the BHSC analytical subcommittee and the laboratory directors at the various sites, who provided the questionnaire responses. We also appreciate the valuable assistance of John R. Longo, Savannah River Site, who compiled the responses into a useful electronic format.

---

[7] The authors do not necessarily endorse the conclusions reached in this comparison.

References

[1] About Beryllium, Department of Energy Chronic Beryllium Disease Prevention Program, URL: http://www.eh.doe.gov, U. S. Department of Energy, 21 January 2005.
[2] Kreiss, K., Mroz, M.M., Zhen, B., Marrtny, J.W., and Newman, L.S., "Epidemiology of Beryllium Sensitization and Disease in Nuclear Workers," *American Review of Respiratory Disease,* Vol. 148, pp. 985-991 (1993).
[3] Tinkle, S.S., Antonini, J.M., Rich, B.A., Roberts, J.R., Salmen, R., DePree, K., Adkins, E.J., "Skin as a Route of Exposure and Sensitization in Chronic Beryllium Disease," *Environmental Health Perspectives,* Vol. 111, pp. 1202-1208 (2003).
[4] *Code of Federal Regulations,* Title 10, Part 850.
[5] Kerr, K., "Sampling Beryllium Surface Contamination Using Wet, Dry, and Alcohol Wipe Methods," M.S. Thesis, Central Missouri State University, Warrensburg, MO (2004).
[6] NIOSH Method 7300, Issue 3, "Elements by ICP," in *NIOSH Manual of Analytical Methods,* Fourth Edition. NIOSH, Cincinnati, OH (1994), URL: http://www.cdc.gov/niosh/nmam, U.S. Centers for Disease Control and Prevention, 24 January 2005.
[7] National Institute of Standards and Technology (NIST), *Standard Reference Materials (SRM) Catalog*, National Institute of Standards and Technology, Gaithersburg, MD; URL: http://ts.nist.gov, 24 January 2005.
[8] U.S. EPA Method 200.7, "Metals and Trace Elements – ICP-AES," EPA Office of Solid Waste, Washington, DC (1994).
[9] U.S. EPA, Method 200.8, "Trace Elements in Water and Wastes – ICP-MS," EPA Office of Solid Waste, Washington, DC (1994).
[10]Cable-Dunlap, P., Guanlao, A., Kahal, E., LaMont, S., Shick, C., and Jahn, S. "Beryllium Wipe Analysis – SRS Lessons Learned," Savannah River Site Memo SRT-NTS-2003-00088, as accessed from U.S. DOE Lessons Learned Database, URL: http://www.eh.doe.gov/ll/images/sitell/ll03/be_attach.pdf, 24 January 2005.
[11] Environmental Express Web Site, URL: http://www.envexp.com, 24 January 2005.
[12] NIOSH Method 9100, Issue 1, "Lead in Surface Wipe Samples," in *NIOSH Manual of Analytical Methods,* Fourth Edition. NIOSH, Cincinnati, OH (1994), URL: http://www.cdc.gov/niosh/nmam, U.S. Centers for Disease Control and Prevention, 24 January 2005.
[13] U.S. HUD, "HUD Technical Guidelines for the Evaluation and Control of Lead-Based Paint Hazards in Housing". HUD, Washington, DC (1995), URL: http://www.hud.gov/offices/lead/guidelines, 24 January 2005.
[14] U.S. OSHA, Method ID-125G, "Metal and Metalloid Particulates in Workplace Atmospheres (ICP Analysis)." OSHA, Salt Lake City, UT (2003), URL: http://www.osha.gov/dts/stlc/methods, 24 January 2005.
[15] NIOSH Method 9102, Issue 1, "Elements on Wipes," in *NIOSH Manual of Analytical Methods,* Fourth Edition. NIOSH, Cincinnati, OH (2003), URL: http://www.cdc.gov/niosh/nmam, U.S. Centers for Disease Control and Prevention, 24 January 2005.
[16] Ashley, K., Brisson, M.J., and Jahn, S.D., "Standard Methods for Beryllium Sampling and Analysis: Availabilities and Needs," *Journal of ASTM International*, Vol. 2, No. 9, October 2004.
[17] ACGIH, *2004 Threshold Limit Values for Chemical Substances and Physical Agents & Biological Exposure Indices.* ACGIH: Cincinnati, OH (2004).
[18] Creek, K.L, and Brisson, M.J., "Symposium on Beryllium: Sampling and Analysis Beryllium Advanced Technology Assessment Team: A Final Report," *Journal of ASTM International* (symposium 9027, paper #13159)
[19] U.S. EPA, Method 6010B, "Inductively Coupled Plasma – Atomic Emission Spectrometry," in publication SW-846, *Test Methods for Evaluating Solid Waste, Physical/Chemical Methods*. EPA Office of Solid Waste, Washington, DC (1996), URL:

http://www.epa.gov/epaoswer/hazwaste/test/sw846.htm.

[20] U.S. EPA, Method 6020, "Inductively Coupled Plasma – Mass Spectrometry," in publication SW-846, *Test Methods for Evaluating Solid Waste, Physical/Chemical Methods*. EPA Office of Solid Waste, Washington, DC (1994), URL: http://www.epa.gov/epaoswer/hazwaste/test/sw846.htm.

[21] NIOSH Method 7102, Issue 2, "Beryllium and Compounds, as Be," in *NIOSH Manual of Analytical Methods*, Fourth Edition. NIOSH, Cincinnati, OH (2003), URL: http://www.cdc.gov/niosh/nmam, U.S. Centers for Disease Control and Prevention, 24 January 2005.

[22] U.S. EPA, "Technical Support Document for the Assessment of Detection and Quantitation Approaches," Report No. EPA-821-R-03-005. EPA Office of Water, Washington, DC (2003).

[23] Department of Energy (DOE) Technical Standards Program, URL: http://www.eh.doe.gov/techstds, U.S. Department of Energy, 2 February 2005.

[24] Beryllium Health and Safety Committee Web Site, URL: http://www.sandia.gov/BHSC, 24 January 2005.

Journal of ASTM International, October 2005, Vol. 2, No. 9
Paper ID JAI13169
Available online at www.astm.org

Kevin Ashley,[1] *Michael J. Brisson,*[2] *and Steven D. Jahn*[2]

Standard Methods for Beryllium Sampling and Analysis: Availabilities and Needs[3]

ABSTRACT: Conformity in methods for sampling and analysis of beryllium in workplace air and on surfaces is desired, but inconsistencies in sampling and analytical practices often occur among industrial hygienists and laboratory personnel. In an effort to address these issues, this paper gives an overview of standardized methods for sampling and analysis of beryllium in the workplace. A number of published methods is currently available to the industrial hygiene and analytical community, but shortfalls in the use of standardized practices require attention. Also, questions remain concerning the performance of some of the sampling and analytical methodologies that have been promulgated. We attempt to identify needs for new or improved standard sampling protocols, sample preparation techniques, analytical methods, and reference materials. Where applicable, performance data are summarized for standardized methods that are either published or are under development. These include not only ASTM and ISO international standards, but also methods published by government agencies in the USA and abroad. Significant gaps in standard methods and requirements for reference materials remain. For example, consistent practices are lacking for: (a) surface sampling of beryllium in dust; (b) extraction of beryllium from surface dust samples prior to instrumental analysis; and (c) reference materials containing beryllium oxide (especially high-fired BeO). These and other gaps will be identified and shortcomings addressed. An ultimate goal is to provide standard methods which will ensure comparability of data obtained from different sites around the globe.

KEYWORDS: aerosols, analysis, beryllium, reference materials, sample preparation, sampling, standards, surfaces, workplace

Introduction

The health of workers in a number of industries and activities is at risk through occupational exposure to beryllium. In order to estimate workers' exposures to beryllium, occupational contact with airborne and dermal sources of beryllium is typically monitored by sampling and analyzing workplace atmospheres and surfaces, respectively. Workplace air monitoring is carried out because in occupational settings inhalation is ordinarily the most likely route of entry of hazardous substances, such as beryllium, into the body. Dermal contact is another important potential route of occupational exposure and possible sensitization to beryllium. Thus, procedures for measuring surface contamination of beryllium in workplaces are also required.

Conformity in methods for sampling and analysis of beryllium in workplace air and on surfaces is desired, but inconsistencies in sampling and analytical practices often occur among

Manuscript received 11 February 2005; accepted for publication 21 March 2005; published October 2005.
Presented at ASTM Symposium on Beryllium: Sampling and Analysis on 21-22 April 2005 in Reno, NV.
[1] U.S. Department of Health and Human Services, Centers for Disease Control and Prevention, National Institute for Occupational Safety and Health, 4676 Columbia Parkway, Mail Stop R-7, Cincinnati, OH 45226-1998, USA; tel.+1(513)841-4402; fax +1(513)841-4500; e-mail: KAshley@cdc.gov.
[2] Westinghouse Savannah River Company, Savannah River Site, 707-F, Aiken, SC 29808, USA.

industrial hygienists and laboratory personnel. If sampling and analytical methods are not standardized, analytical results from different investigators, locations, or points in time may not be comparable. Variations in sampling practices are of special concern, since the greatest contribution to measurement uncertainty in the overall analysis is ordinarily associated with sampling. Efforts to minimize measurement uncertainty through method standardization have been realized for many workplace toxins and, as a consequence, standard methods for numerous hazardous substances in workplace atmospheres have been promulgated [1]. However, gaps remain in a number of areas where standard techniques are as yet unavailable. For example, surface sampling techniques have not been adequately standardized or, where such standards exist, they have not been put to use in the relevant areas of application. Also, reference materials for beryllium are few and limited. In an effort to address these requirements for new standards, we have endeavored to identify and suggest correction of situations where standard techniques are needed.

Another matter of concern is where existing standards may not meet desired performance criteria or may not be applicable to a given area of application. Questions concerning method performance might only be answered following additional research, but this requires resources (human and capital) that are often unavailable. So in the absence of research results, standards are often produced by consensus with the best intentions in mind. Yet shortcomings can appear when problems in the application of sub-par standards are realized and identified. Thus, in consideration of possible shortcomings of existing standards, we have made an effort to identify problematic standards with a view to improving them or, where necessary, producing new standards that will hopefully be fit for purpose.

In this paper, standards for sampling, sample preparation, analysis, and reference materials for the determination of beryllium in workplace environments are at issue. For each instance we have made an effort to identify existing standards and to provide performance data, where available. In cases where available standards are found to be lacking, the shortcomings have been pointed out, with an ultimate intention of filling these gaps. It is intended that this article will provide potential users of standards relating to sampling and analysis of beryllium in occupational settings with a useful body of information which will enable them to solve their industrial hygiene and analytical chemistry problems.

Sampling

Collection of Workplace Air Samples

Sampling of beryllium and other metals in workplace atmospheres using filter samplers has been well standardized. In the United States, governmental methods have been published by the National Institute for Occupational Safety and Health (NIOSH) [2] and the Occupational Safety and Health Administration (OSHA) [3]. Workplace air sampling entails personal monitoring using portable, battery-powered sampling pumps that pull air at desired flow rates through the samplers, which are placed within the workers' personal breathing zones. Area sampling is also possible, but workers' exposures are best estimated by personal sampling [4]. The NIOSH and OSHA methods describe "total" air samplers consisting of closed-faced sampling cassettes housing mixed-cellulose ester (MCE) membrane filters. These methods have been widely used worldwide for both regulatory and non-regulatory (e.g., research and hazard assessment) purposes.

International voluntary consensus standards describing workplace air monitoring for metals (including beryllium) have been published recently, which are based, at least in part, on the NIOSH and OSHA methods. Procedures promulgated by the International Organization for Standardization (ISO) [5] and ASTM International [6] describe sampling of the inhalable fraction of atmospheric aerosols using filter samplers and personal sampling pumps. Performance-based options in these consensus standards also allow for sampling of the respirable aerosol fraction, the use of various sampler designs, and the use of fibrous or other membranous filters (besides MCE), depending on the intended application. Some of the technical options for sampling of workplace aerosols that are described in the ISO and ASTM International standards, which differ significantly from the NIOSH and OSHA procedures, have appeared in governmental methods published in France [7], Germany [8], and the United Kingdom [9].

Table 1 summarizes several governmental and two new consensus standard sampling methods for beryllium in workplace atmospheres. Collection efficiencies of 99.5 % or better are afforded by use of these air sampling methods. Generally, the methods listed in Table 1 recommend full 8-h time-weighted average (TWA) personal sampling, but shorter sampling periods can also be employed for certain applications, such as task-based workplace exposure monitoring [10]. The sampling procedures listed in Table 1 allow for subsequent determination of trace quantities of collected airborne beryllium.

TABLE 1—*Governmental and consensus standard procedures for sampling beryllium aerosols in workplace atmospheres.*

Method(s)	Aerosol Fraction(s) Sampled	Filter Type(s) Used
NIOSH 7102 & 7300	"Total"	Membrane
OSHA ID-125G & ID-206	"Total"	Membrane
HSE 29/2 (UK)	Inhalable	Membrane
BIA 6015 (Germany)	Inhalable	Membrane
INRS Fiche 003 (France)	Inhalable	Quartz Fiber
ISO 15202-1	Inhalable or Respirable	Membrane or Fibrous
ASTM D7035	"Total," Inhalable or Respirable	Membrane or Fibrous

Collection of Surface Samples

Workers can become sensitized to beryllium through dermal contact with beryllium-containing dust, and this can ultimately lead to chronic beryllium disease [11]. In order to assess contamination and prevent dermal contact with beryllium-contaminated surfaces, methods are desired for collection of surface dust samples for subsequent determination of this highly toxic metal. For maximum collection efficiency, "wet" sampling techniques using wipes are generally preferred. However, there are instances where wet sampling of certain components and equipment are not desirable and dry sampling techniques are required. As an example, for technical reasons, surfaces of certain materials and components must be protected against damage from sample collection; hence, sampling methods that are less aggressive than wipe sampling are required. As shown in Table 2, several standardized procedures for wet and dry sampling have been produced by various organizations. However, standardized sampling methods have not been applied uniformly, and this has led to difficulties in comparing data obtained from different sites and locations. Also, performance data are often lacking for surface sampling techniques, and, as a consequence, collection efficiencies may not be known. Thus, efforts are underway to characterize surface sampling procedures that have been promulgated by various sources, in order that: (a) comparisons may be made among data from wet and dry sampling procedures; and (b) recommendations can be made on the best sampling techniques for beryllium in surface dust.

TABLE 2—*Standardized sampling procedures for collecting beryllium dust samples from surfaces.*

Method(s)	Sampling Media/Device	Substrate(s) Sampled	Comments
OSHA ID-125G & ID-206	"Wet" or "dry" filter or wipe	Smooth surfaces, dermal samples	Alcohol wipes widely used
NIOSH 9100 & 9102	"Wet" wipe	Smooth surfaces, dermal samples	Individually packaged wipes
ASTM D 6966	"Wet" wipe	Smooth surfaces	Individually packaged wipes; ASTM E 1792 wipes acceptable for Be sampling
ASTM E 1216	Adhesive tape	Smooth surfaces	May damage fragile surfaces
OSHA Technical Manual	Patch samples, hand rinsates	Dermal samples	Various protocols; may sample clothing, gloves, etc.
NIOSH 2600, 3601, 9202 & 9205	Patch samples, hand rinsates	Dermal samples	Developed for sampling pesticides, metalworking fluids, etc.
ASTM D 5438	Modified upright vacuum cleaner	Floors	Sampling from carpets
ASTM E 1973	Sampling cassette with collection nozzle	Rough, porous, uneven surfaces	"microvacuum" Pb dust sampling (withdrawn standard)
ASTM WK4996 (work item)	Sampling cassette with collection nozzle	Rough, porous, uneven surfaces; fragile surfaces	"microvacuum" dust sampling for metals (under development; to replace ASTM E 1973)

On its web page, OSHA has published procedures for obtaining surface samples for assessment of surface contamination [12]. These methods have been recommended for surface sampling for contaminants such as beryllium [3]. The OSHA procedures describe wipe sampling techniques for worksite and dermal surfaces, generally using filters (membrane or fiber, depending on the application) as sampling media. These procedures are currently the only ones published that describe the use of dry sampling media. While filters are easy to handle and their sample preparation and analysis are uncomplicated, the collection efficiency of sampling surface dust when using filters is highly questionable. OSHA has also promulgated evaluation guidelines for surface sampling methods [13], but the procedures described therein prescribe sample collection from ultra-smooth surfaces that are not representative of real-world situations. Due in part to the shortcomings of available surface sampling methods described by OSHA, others have sought to employ and evaluate alternative surface sampling methods for standardization purposes.

Apart from sampling media, there are other confounders in evaluating surface sampling methods. Notably, non-uniformity of surface contamination, e.g., in terms of beryllium loading levels, beryllium compounds, and particle size distribution, can lead to variable results. This is another factor to consider in efforts to harmonize surface sampling methods.

In a recent investigation, a comparison of wipe sampling methods for beryllium was carried out wherein dry, wet, and alcohol wipe methods were evaluated for their application in removing beryllium-containing dust from painted surfaces [14]. This study found alcohol to be most effective for removing beryllium dust from oily surfaces, while (not surprisingly) dry wipes were least effective for this purpose. Wipe sampling of surfaces using robust wipe media, wetted with either distilled water or an organic solvent (such as methanol), has been described in a standard practice developed by ASTM International [15]. This consensus standard, ASTM D 6966, describes the collection of surface wipe samples from delineated areas of smooth, flat surfaces from components such as furniture, room components, and equipment encountered in worksites. The ASTM D 6966 wipe sampling procedure is based on a NIOSH method for collection of surface dust samples for subsequent determination of lead [16]. NIOSH recently extended the wipe sampling procedure for lead to multi-element determination [17]. Candidate wipe materials for sampling may include wipes meeting the specifications of ASTM E 1792 [18], which was developed to address wipe materials for lead dust collection. But besides the results from sampling of oily surfaces presented in the aforementioned study [14], general performance data for beryllium dust sampling from representative surfaces when using ASTM D 6966 are as yet unavailable.

Another consensus standard, ASTM E 1216, describes the collection of surface contaminants in particulate matter using pressure-sensitive tape [19]. This standard practice is targeted for use on surfaces that are not damaged by the use of adhesive tape for sampling purposes. It is meant for the collection of particles 5 μm in diameter and larger, but does not efficiently collect ultrafine (<1 μm) beryllium particulate matter, which is of concern [20]. ASTM E 1216 is generally applicable to surface sampling from metal platings, oxide coatings, and robust materials. But the use of this method on painted, vapor-deposited, and optical coatings is recommended with caution and should be carried out only after preliminary evaluation.

Collection of dermal samples from workers' hands has been described briefly in a new NIOSH method for colorimetric lead screening in dust wipe samples [21]. The procedure entails the use of wetted wipes to remove contaminated dust from hands for subsequent analysis. The worker to be sampled is asked to use a fresh wipe to remove dust from first the palms, then the fingers, and lastly, the backs of the hands, with wiping to be done for a period of 30 s per pair of hands. The method is similar to the procedure described by OSHA [12], except that wetted wipes, rather than filters, are specified in the NIOSH 9105 method for dermal (hand) sampling. Other dermal sampling methods that have been used for various skin contaminants, such as patch sampling and collection of hand rinsates, have also been described by NIOSH [22] and OSHA [12]. Unfortunately, dermal sampling procedures have not been well standardized, and this has led to difficulties in evaluating and comparing data from a variety of different studies [23]. Dermal sampling methods for beryllium need to be harmonized, and this remains an important area for further research and development efforts.

Vacuum collection techniques are sometimes used in "dry" methods to sample surface dust, and some of these procedures have been standardized. Sample collection methods using a vacuum may be preferred in lieu of "wet" methods requiring wipes for situations where surfaces are rough or highly porous, or both. For collection of floor dust for subsequent analysis, a standardized vacuum sampling method has been developed by ASTM International [24]. The procedure is applicable to indoor environments including workplaces, but it has not been evaluated for purposes of beryllium sampling and analysis. The method is especially useful for sampling larger areas of uneven substrates, such as carpets. Vacuum sampling might also be required in cases where it is necessary to prevent damage to the substrate being sampled through direct contact with a collection medium, such as a wipe. To this end, a "microvacuum" dust sampling method is under development in ASTM Subcommittee D22.04 on Sampling and Analysis of Workplace Atmospheres, with collected samples intended for subsequent determination of metals, including beryllium [25]. The method is an extension of an earlier (and now withdrawn) consensus standard describing microvacuum sampling for subsequent lead determination [26]. This procedure entails vacuum collection of surface dust using sampling cassettes to which collection nozzles are attached [27]. With vacuum generated by a low-flow sampling pump, the collection nozzle is used to sample settled dust on delineated areas of surfaces of interest. While it is known that only smaller particles will be collected, no damage will befall the substrate when using this technique since the collection nozzle does not touch the substrate being sampled.

Standardized surface sampling methods that can be applied to the collection of beryllium in surface dust are summarized in Table 2. Unfortunately, performance data are mostly lacking for these procedures, and additional research is needed for the assessment of surface dust sampling techniques for the subsequent determination of beryllium. Nevertheless, in the absence of performance figures of merit, the use of harmonized surface sampling methods will enable better comparability of data from different samplers, sampling sites, locations, and times.

Methods for obtaining bulk samples are outside the scope of this article, but they are briefly mentioned here. An excellent source of information on bulk sampling of soils, solid waste, water, field equipment, etc. is the U.S. Environmental Protection Agency (EPA), which has published an exhaustive tome [28] that covers issues such as: (a) sampling strategies and design; (b) sampling techniques, media, and equipment; (c) standardized sampling procedures developed through voluntary consensus (notably ASTM standards); (d) data quality considerations pertaining to sample collection, sample handling, and transport; and (e) numerous related subjects. A great many relevant ASTM standards on collecting bulk samples have also appeared in ASTM publications on environmental sampling [29,30]. Further study is also needed regarding when it is more appropriate to use bulk sampling as opposed to surface sampling for beryllium. Guidelines and performance data are limited in this area.

Sample Preparation

Laboratory methods are required to prepare beryllium samples for subsequent analysis. To this end, several standardized procedures describing beryllium dissolution techniques have been published. Owing to the refractory nature of beryllium particulate matter, including beryllium oxide (especially the high-fired variety) and beryllium silicates, rigorous strong acid high temperature digestion techniques are generally needed in order to dissolve beryllium samples. Sulfuric or hydrofluoric acids (in concert with other acids such as nitric or hydrochloric) are ordinarily required to put refractory beryllium compounds into solution.

Standardized high temperature acid digestion techniques are available for the purpose of dissolution of beryllium in samples collected from workplaces. A number of options for dissolution procedures has been described in recently published international standards [6,31], but performance data for beryllium recoveries from refractory materials have not yet been obtained through interlaboratory testing. Hot plate or microwave digestion techniques are widely recommended, but ultrasonic extraction is an option that can be used with hydrofluoric acid digestion. Various governmental and consensus standard methods describing the preparation of beryllium samples from occupational environments are summarized in Table 3. Procedures for the dissolution of wipe samples require larger volumes of acids than the quantities needed for air filter samples. Also, if microwave digestion of wipe samples is carried out, this may require a preliminary hot plate dissolution step. Unfortunately, methods for preparing wipe samples for beryllium determination have not been well standardized to date. An ASTM standard describing sample preparation methods for wipes is currently being considered for development within ASTM Subcommittee D22.04 on Sampling and Analysis of Workplace Atmospheres.

Acid digestion methods for the preparation of bulk samples for subsequent beryllium determination are generally similar to those for air filters. Besides the OSHA procedures listed in Table 3, there is a number of EPA [28] and ASTM [30] hot plate and microwave acid digestion methods that may be applicable to the dissolution of beryllium in bulk samples collected from workplaces.

TABLE 3—*Governmental and consensus standard procedures for preparation of beryllium samples obtained in workplaces.*

Method	Sample Type(s)	Acid Mixture	Digestion Method(s)
NIOSH 7102	Air filter	Nitric & Sulfuric	Hot plate
NIOSH 7300	Air filter	Nitric & perchloric	Hot plate or microwave
NIOSH 7302	Wipe	Nitric & perchloric	Hot plate or microwave
OSHA ID-125G	Air filter, wipe, or bulk	Nitric, sulfuric, & hydrochloric	Hot plate
OSHA ID-206	Air filter, wipe (smear tab), or bulk	Hydrochloric, nitric	Hot plate
HSE 29/2 (UK)	Air filter	Nitric & Sulfuric	Hot plate
INRS Fiche 003 (France)	Air filter	Nitric & hydrofluoric	Hot plate or microwave
ASTM D7035	Air filter	Various options	Hot plate or microwave
ISO 15202-2	Air filter	Various options	Hot plate, microwave, or ultrasound

Laboratory Analysis

Following sample dissolution, laboratory analytical methods are then used to measure the concentrations of dissolved beryllium in sample extracts. Standardized instrumental analytical methods for beryllium are generally based on atomic spectrometric techniques. Graphite furnace atomic absorbance spectrometry (GFAAS) and inductively coupled plasma atomic emission spectrometry (ICP-AES) are the two most widely used instrumental methods for determining beryllium in extracts from workplace samples. Atomic spectrometric methods for beryllium, as evaluated by NIOSH [2] and OSHA [3], are very precise and unbiased.

New ICP-AES voluntary consensus standards, which were in development for nearly a decade, have recently been published [6,32]; these international standard methods are applicable to workplace beryllium measurements. If ultra-trace beryllium determinations are required, inductively coupled plasma mass spectrometry (ICP-MS) can be employed. EPA methods for determining beryllium in environmental samples based on GFAAS [28,33], ICP-AES [28,34], or ICP-MS [28,35] have been published. Methods using flame atomic absorption spectrometry (FAAS) are also available [28,36], but method detection limits for FAAS are generally inadequate for the trace analyses that are ordinarily required in workplace beryllium monitoring.

Several standardized analytical methods that can be used to determine beryllium in workplace samples are summarized in Table 4. All of these methods entail aspiration of aliquots of sample extracts into the instrument. GFAAS and ICP-AES yield comparable method detection limits for beryllium. ICP-AES is especially useful for analysis since other elements besides beryllium can also be monitored simultaneously. For cost considerations, ICP-MS is generally not recommended for routine sample analysis. But this methodology may be necessary for determining beryllium in, for example, short-term task-based workplace air samples, or in other instances where very low detection limits may be required. All method detection limits for the methods listed in Table 4 are well below the action level of 0.2 μg Be per sample established for air [37] (1000-L sampling volume) and surface [38] (100 cm^2) samples.

TABLE 4—*Governmental and consensus standard methods for atomic spectrometric analysis of workplace beryllium samples.*

Method	Instrumental Technique	Estimated Method Detection Limit (μg Be/sample)
NIOSH 7102	GFAAS	0.005
NIOSH 7300	ICP-AES	0.005
OSHA ID-125G	ICP-AES	0.013
OSHA ID-206	ICP-AES	0.0072
EPA 7091	GFAAS	0.005
EPA 200.7	ICP-AES	0.008
EPA 200.8	ICP-MS	0.001
EPA 6010B	ICP-AES	0.005*
EPA 6020	ICP-MS	0.0005*
ASTM D7035	ICP-AES	0.009
ISO 15202-3	ICP-AES	Not evaluated

*(Based on a 25-mL sample).

Reference Materials

To evaluate sampling and analytical methods, representative certified reference materials containing beryllium are desired. Several National Institute of Standards and Technology (NIST) Standard Reference Materials (SRMs) having certified reference concentrations of beryllium are available [39]; these are summarized in Table 5. Copper-beryllium alloy SRMs are well represented; these are in the form of chips (≈0.5 to ≈1.2 mm diameter for SRMs 458, 459, and 460) or blocks (19 mm × 31 mm for SRM C1122). Highly divided powders containing certified concentrations of beryllium are represented by bituminous coal (SRM 1632c) and waterway sediment (SRM 1944). Certified beryllium in solution is available as a single element standard solution (SRM 3105a) and as an isotope standard for applications in mass spectrometry (SRM 4325). Some older Be-Cu alloy SRMs have been discontinued, as has an old SRM consisting of beryllium and arsenic spiked onto filter media. There are several other SRMs containing beryllium at non-certified concentrations [39].

TABLE 5—*Available NIST SRMs containing beryllium at certified concentrations.*

SRM #	Description	Certified Beryllium Concentration
458	Be-Cu alloy (chips)	0.360 % (by weight)
459	Be-Cu alloy (chips)	1.82 % (by weight)
460	Be-Cu alloy (chips)	1.86 % (by weight)
C1122	Copper base alloy (block)	1.75 % (by weight)
1632c	Coal (bituminous)	1.0 µg/g
1944	NY/NJ waterway sediment	1.6 µg/g
3105a	Single element standard solution	10 mg/L
4325	Be 10/9 accelerator mass spectrometry standard (solution form)	5 mg/L; Be 10/9 ratio = 3×10^{-11}

It can be seen from the list of Table 5 that beryllium-containing SRMs are unavailable for media such as air filters and wipes. Also, environmental SRMs containing beryllium oxide are not available either. Efforts are presently underway by the National Institute of Standards and Technology (NIST) to develop representative SRMs containing beryllium oxide, including the high-fired variety, which is particularly refractory in nature. Additionally, there remains a need to certify beryllium concentrations in representative environmental matrices.

Summary

Various standardized sampling, sample preparation, and analytical methods for beryllium in workplace samples have been published by governmental agencies and consensus standards organizations. Within the beryllium analysis arena, our goal is to encourage the development of voluntary consensus standards in areas of interest for which such standards are presently unavailable. Methods for measuring beryllium in workplace air samples are well standardized, as evidenced by the availability of recently published ASTM and ISO international standards. However, taken as a whole, surface sampling methods for beryllium require better harmonization and evaluation. A new ASTM standard procedure for obtaining surface dust wipe samples using wet wiping is available, and, for collecting "dry" samples, a "microvacuum" ASTM standard sampling method is under development. A vacuum sampling method for collecting floor dust has also been standardized in the form of an ASTM procedure. There remains a need for a voluntary consensus standard to describe sample preparation procedures for beryllium in wipe samples. Also, for many of the existing standard procedures, performance data are lacking for refractory beryllium compounds. Reference materials containing beryllium oxide, as well as beryllium in several environmental matrices of interest, are unavailable to date. Candidate areas for further standardization in the beryllium sampling and analysis arena will be undertaken when needs for new standards are identified.

Acknowledgments

This work was carried out in coordination with the Analytical Subcommittee of the Beryllium Health and Safety Committee [40]. We thank Jensen Groff, Eugene Kennedy, Rosa Key-Schwartz, and the referees for their critical review of the draft manuscript.

References

[1] *Annual Book of ASTM Standards*, Vol. 11.03, ASTM International, West Conshohocken, PA, 2004 (updated annually), www.astm.org (accessed 6 Jan. 2005).

[2] U.S. National Institute for Occupational Safety and Health (NIOSH), *NIOSH Manual of Analytical Methods*, 4th ed., Method Nos. 7102 & 7300, NIOSH, Cincinnati, OH, 1994, www.cdc.gov/niosh/nmam (accessed 3 Jan. 2005).

[3] U.S. Occupational Safety and Health Administration (OSHA), *OSHA Sampling and Analytical Methods*, Method Nos. ID-125G & ID-206, OSHA, Salt Lake City, UT, 2003, www.osha.gov/dts/sltc/methods (accessed 3 Jan 2005).

[4] Lippman, M., "Sampling Aerosols by Filtration," *Air Sampling Instruments*, 7th ed., S. V. Hering, Ed., American Conference of Governmental Industrial Hygienists (ACGIH), Cincinnati, OH, 1989, pp. 305–336.

[5] ISO 15202-1, *Workplace Air – Determination of Metals and Metalloids in Airborne Particulate Matter by Inductively Coupled Plasma Atomic Emission Spectrometry – Part 1: Sampling*, International Organization for Standardization (ISO), Geneva, Switzerland, 2000.

[6] ASTM Standard D 7035, "Standard Test Method for the Determination of Metals and Metalloids by Inductively Coupled Plasma Atomic Emission Spectrometry," *Annual Book of ASTM Standards*, ASTM International, West Conshohocken, PA, 2004.

[7] Institut National de Recherche et de Sécurité (INRS), *Métrologie des Polluants – Evaluation de l'Exposition Professionnelle – Méthodes de Prélèvement et d'Analyse de l'Air*, Fiche 003, INRS, Paris, France 2004, (updated annually), www.inrs.fr (accessed 5 Jan. 2005).

[8] Berufsgenossenschaftlisches Institut für Arbeitssicherheit (BIA), *Measurement of Hazardous Substances in Air – Determination of Exposure to Chemical and Biological Agents*, Method 6105, BIA, Sankt Augustin, Germany, 1989.

[9] U.K. Health and Safety Executive (HSE), *Methods for the Determination of Hazardous Substances*, MDHS Method No. 29/2. HSE Books: London, England, 1996, www.hsebooks.co.uk (accessed 5 Jan. 2005).

[10] ASTM Standard E 1370, "Standard Guide for Air Sampling Strategies for Worker and Workplace Protection," *Annual Book of ASTM Standards*, ASTM International, West Conshohocken, PA, 1996, revised 2002.

[11] Tinkle, S. S., Antonini, J. M., Rich, B. A., Roberts, J. R., Salmen, R., DePree, K., et al., "Skin as a Route of Exposure and Sensitization in Chronic Beryllium Disease," *Environmental Health Perspectives*, Vol. 111, 2003, pp. 1202–1208.

[12] OSHA, "Sampling for Surface Contamination," *OSHA Technical Manual*, Section II, Ch. 2, OSHA, Washington, DC, 1999, www.osha.gov (accessed 6 Jan. 2005).

[13] OSHA, "Evaluation Guidelines for Surface Sampling Methods," in *OSHA Sampling and Analytical Methods*, OSHA, Salt Lake City, UT, 2001, www.osha.gov/dts/sltc/methods/surfacesampling (accessed 6 Jan. 2005).

[14] Kerr, K., "Sampling Beryllium Surface Contamination Using Wet, Dry and Alcohol Wipe Methods," M.S. Thesis, Central Missouri State University, Warrensburg, MO, 2004.

[15] ASTM Standard D 6966, "Standard Practice for Collection of Surface Dust Using Wipe Sampling Method for Subsequent Determination of Metals," *Annual Book of ASTM Standards*, ASTM International, West Conshohocken, PA, 2003.

[16] NIOSH Method 9100, "Lead in Surface Dust Wipe Samples," *NIOSH Manual of Analytical Methods*, 4th ed., 1st Suppl., NIOSH, Cincinnati, OH, 1996, www.cdc.gov/niosh/nmam (accessed 7 Jan. 2005).

[17] NIOSH Method 9102, "Elements on Wipes," *NIOSH Manual of Analytical Methods*, 4th ed., 3rd Suppl., NIOSH, Cincinnati, OH, 2004, www.cdc.gov/niosh/nmam (accessed 7 Jan. 2005).

[18] ASTM Standard E 1792, "Standard Specification for Wipe Sampling Materials for Lead in Surface Dust," *Annual Book of ASTM Standards*, ASTM International: West Conshohocken, PA, 2003.

[19] ASTM Standard E 1216, "Standard Practice for Sampling for Particulate Contamination by Tape Lift," *Annual Book of ASTM Standards*, ASTM International: West Conshohocken, PA, 1999.

[20] Stefaniak, A., Hoover, M. D., Day, G. A., Dickerson, R. M., Peterson, E. J., Kent, M. S., et al., "Characterization of Physicochemical Properties of Beryllium Aerosols Associated with Prevalence of Chronic Beryllium Disease," *Journal of Environmental Monitoring*, Vol. 6, 2004, pp. 523–532.

[21] NIOSH Method 9105, "Lead in Dust Wipes by Chemical Spot Test," *NIOSH Manual of Analytical Methods*, 4th ed., 3rd Suppl., NIOSH, Cincinnati, OH, 2004, www.cdc.gov/niosh/nmam (accessed 7 Jan. 2005).

[22] *NIOSH Manual of Analytical Methods*, 4th ed., 3rd Suppl., Method Nos. 2600, 3601, 9202, 9205, NIOSH, Cincinnati, OH, 1994 & 2004, www.cdc.gov/niosh/nmam (accessed 7 Jan. 2005).

[23] Brouwer, D. H., Boeniger, M. F., and Van Hemmen, J., "Hand Wash and Manual Skin Wipes," *Annals of Occupational Hygiene*, Vol. 44, 2000, pp. 501–510.

[24] ASTM Standard D 5438, "Standard Practice for Collection of Floor Dust for Chemical Analysis," *Annual Book of ASTM Standards*, ASTM International, West Conshohocken, PA, 2000.

[25] ASTM WK 4996, "Practice for Collection of Surface Dust by Microvacuum Technique for Subsequent Determination of Metals," Draft Work Item, ASTM International: West Conshohocken, PA, 2005.

[26] ASTM Standard E 1973, "Standard Practice for Collection of Surface Dust by Air Sampling Pump Vacuum Technique for Subsequent Lead Determination," *Annual Book of ASTM Standards*, ASTM International, West Conshohocken, PA, 1999, withdrawn 2004.

[27] Que Hee, S. S., Peace, B., Clark, C. S., Boyle, J. R., Bornschein, R. L., and Hammond, P. B., "Evolution of Efficient Methods to Sample Lead Sources, Such as House Dust and Hand Dust, in the Homes of Children," *Environmental Research*, Vol. 38, 1985, pp. 77–95.

[28] U.S. Environmental Protection Agency (EPA), *RCRA Waste Sampling Draft Technical Guidance – Planning, Implementation and Assessment*, (EPA 530-D-02-002), EPA Office of Solid Waste, Washington, DC, 2002, www.epa.gov/osw (accessed 11 Jan. 2005).

[29] *ASTM Standards on Environmental Sampling*, 2nd ed., ASTM International, West Conshohocken, PA, 1997.

[30] *ASTM Standards on Environmental Site Characterization*, 2nd ed., ASTM International, West Conshohocken, PA, 2002.

[31] ISO 15202-2, *Workplace Air – Determination of Metals and Metalloids in Airborne Particulate Matter by Inductively Coupled Plasma Atomic Emission Spectrometry – Part 2:*

Sample Preparation, International Organization for Standardization (ISO), Geneva, Switzerland, 2001.

[32] ISO 15202-3, *Workplace Air – Determination of Metals and Metalloids in Airborne Particulate Matter by Inductively Coupled Plasma Atomic Emission Spectrometry – Part 3: Analysis*, International Organization for Standardization (ISO), Geneva, Switzerland, 2005.

[33] EPA, "Beryllium, Atomic Absorption, Furnace Technique," (Method No. 7091), EPA Office of Solid Waste, Washington, DC, 1986, www.epa.gov/epaoswer (accessed 12 Jan. 2005).

[34] EPA, "Metals and Trace Elements – ICP-AES," (Method No. 200.7), EPA Office of Solid Waste, Washington, DC, 1994.

[35] EPA, "Trace Elements in Water & Wastes – ICP-MS," (Method No. 200.8), EPA Office of Solid Waste, Washington, DC, 1994.

[36] EPA, "Beryllium, Atomic Absorption," (Method No. 7090), EPA Office of Solid Waste, Washington, DC, 1986, www.epa.gov/epaoswer (accessed 12 Jan. 2005).

[37] American Conference of Governmental Industrial Hygienists, *2004 Threshold Limit Values for Chemical Substances and Physical Agents & Biological Exposure Indices*, ACGIH, Cincinnati, OH, 2004, updated annually.

[38] Code of Federal Regulations, 10 CFR Part 850, *Chronic Beryllium Disease Prevention Program*, U.S. Department of Energy, Washington, DC, 1999.

[39] National Institute of Standards and Technology (NIST), *Standard Reference Materials (SRM) Catalog*, NIST, Gaithersburg, MD, 2004, http://ts.nist.gov (accessed 13 Jan. 2005).

[40] www.sandia.gov/BHSC/subs/analytical.htm (accessed 10 Dec. 2004).

BERYLLIUM EXPOSURE MEASUREMENT AND REFERENCE MATERIALS—NATIONAL AND INTERNATIONAL PERSPECTIVES

Journal of ASTM International, Jan. 2006, Vol. 3, No. 1
Paper ID: JAI13171
Available online at: www.astm.org

Robert L. Watters, Jr.,[1] *Mark D. Hoover,*[2] *Gregory A. Day,*[2] *and Aleksandr B. Stefaniak*[2]

Opportunities for Development of Reference Materials for Beryllium

ABSTRACT: Reference materials provide the foundation for assessment of analytical chemistry methods, accurate quantification of occupational and environmental exposures, and conduct of in vitro and in vivo toxicology studies for health effects research. Although the National Institute of Standards and Technology (NIST) supplies industry, academia, government, and other users with over 1300 reference materials of the highest quality and metrological value, the number of beryllium reference materials is limited. Currently available beryllium reference materials include standard spectroscopy solutions of beryllium and copper-beryllium alloy in the form of blocks, chips, and rods. Beryllium is present as a trace element in some standard soil-sludge, coal fly ash, and urine reference materials. Beryllium on filter media was available at one time, but is not currently available. A number of opportunities exist for identification and development of needed beryllium reference materials for beryllium-containing ores, beryllium oxide, beryllium metal, beryllium metal-matrix materials, beryllium-containing alloys, and beryllium in biological samples. These opportunities will require multi-disciplinary and multi-organizational collaboration. Needed actions include consensus on the relevant chemical and physical forms of beryllium; market analyses of demand for the materials; identification of candidate industrial or laboratory-produced samples of the materials; selection of samples that meet criteria for uniformity, physical form, measured quantities, and continued availability; development of production protocols for collection and preparation of the materials, including adequate provisions for occupational health and environmental protection; incorporation of these materials into a supply, distribution, and cost-recovery infrastructure; and continued feedback and information sharing to ensure that the reference materials are meeting user needs or are modified as necessary. Lessons from other major initiatives for reference materials of lead, silica, and similar materials provide guidance on how to optimize and implement an enhanced program for beryllium reference materials.

KEYWORDS: beryllium, reference materials, certified reference materials, standard reference materials, traceability

Nomenclature

Certified Reference Material (CRM)—Reference material, accompanied by a certificate, one or more of whose property values are certified by a procedure which establishes its traceability to an accurate realization of the unit in which the property values are expressed, and for which each certified value is accompanied by an uncertainty at a stated level of confidence [1].

Measurand—Particular quantity subject to measurement [2].

Reference Material (RM)—Material or substance one or more of whose property values are sufficiently homogeneous, stable, and well established to be used for the calibration of an apparatus, the assessment of a measurement method, or for assigning values to materials [1].

Manuscript received April 14, 2005; accepted for publication June 30, 2005; published Jan. 2006.
[1] Chief, Measurement Services Division, National Institute of Standards and Technology, Gaithersburg, Maryland 20899-2320.
[2] Industrial Hygienist, National Institute for Occupational Safety and Health, Morgantown, West Virginia 26505-2888.

Standard Reference Material® (SRM)—A certified reference material (CRM) issued by the National Institute of Standards and Technology (NIST). An SRM is a well characterized material produced in quantity to improve measurement science. It is certified for specific chemical or physical properties, and is issued by NIST with a certificate that reports the results of the characterization and indicates the intended use of the material.

Traceability—The property of the result of a measurement or the value of a standard whereby it can be related to stated references, usually national or international standards, through an unbroken chain of comparisons all having stated uncertainties [2].

Introduction

Reference materials play a critical role in occupational health efforts to understand and prevent disease from toxic materials such as beryllium. Reference materials provide the foundation for assessment of analytical chemistry methods, for accurate quantification of occupational and environmental exposures, and for the conduct of empirical and mechanistic health effects research. Reference materials are also critical components of material science, engineering, and production.

Although the National Institute of Standards and Technology (NIST) supplies industry, academia, government, and other users with over 1300 reference materials of the highest quality and metrological value, the number of beryllium materials is limited. Currently available beryllium reference materials include standard spectroscopy solutions of beryllium and copper-beryllium in the form of blocks, chips, and rods. Beryllium is also present as a trace element in some standard soil-sludge, coal fly ash, and urine reference materials. Beryllium on filter media was available at one time, but is not currently available.

A number of opportunities exist for identification and development of needed beryllium reference materials for beryllium-containing ores, beryllium oxide, beryllium metal, beryllium metal-matrix materials, beryllium-containing alloys, and beryllium in biological samples. These opportunities will require multi-disciplinary and multi-organizational collaboration. The purpose of this paper is to summarize the terminology and qualification procedures for reference materials, present examples of relevant reference materials, and discuss the possible next steps for development of new beryllium reference materials.

Types of Reference Materials

A *Reference Material (RM)* is a material or substance one or more of whose property values are sufficiently homogeneous, stable, and well established to be used for the calibration of an apparatus, the assessment of a measurement method, or for assigning values to materials [1]. Figure 1 illustrates the hierarchy among all materials in our workplaces and general environment and the subsets of materials that qualify as various types of reference materials.

A "material of interest" becomes a reference material by being selected and prepared in an appropriate manner:

$$\begin{pmatrix}\text{A material}\\\text{of interest}\end{pmatrix} + \begin{pmatrix}\text{Adequate Preparation}\\\text{(homogeneity, uniformity, etc)}\end{pmatrix} = \begin{pmatrix}\text{A Reference}\\\text{Material (RM)}\end{pmatrix}$$

A reference material qualifies as a *Certified Reference Material (CRM)* when a metrology laboratory (national or commercial) issues a certificate stating that one or more of its property values are certified by a procedure which establishes its traceability to an accurate realization of the unit in which the property values are expressed, and for which each certified value is accompanied by an uncertainty at a stated level of confidence [1]. Thus:

$$\begin{pmatrix}\text{A Reference}\\\text{Material (RM)}\end{pmatrix} + \begin{pmatrix}\text{Certified}\\\text{measurand}\\\text{values}\end{pmatrix} = \begin{pmatrix}\text{A Certified}\\\text{Reference}\\\text{Material}\\\text{(CRM)}\end{pmatrix}$$

The differences between an RM and a CRM are that (1) the CRM requires metrological efforts by the producer to ensure the traceability of the certified results for the CRM and (2) the CRM requires the issuance of a certificate by that producer taking responsibility for the reported values. To conform to an accreditation program or to gain wider acceptance, CRM producers often adhere to the general requirements for the competence of reference material producers as stated in ISO Guide 34 [3] and to the general and statistical principles for the certification of reference materials as stated in ISO 35 [4], and ensure that their certificates fulfill the requirements of ISO Guide 31 [5], which defines the contents of certificates and labels for reference materials.

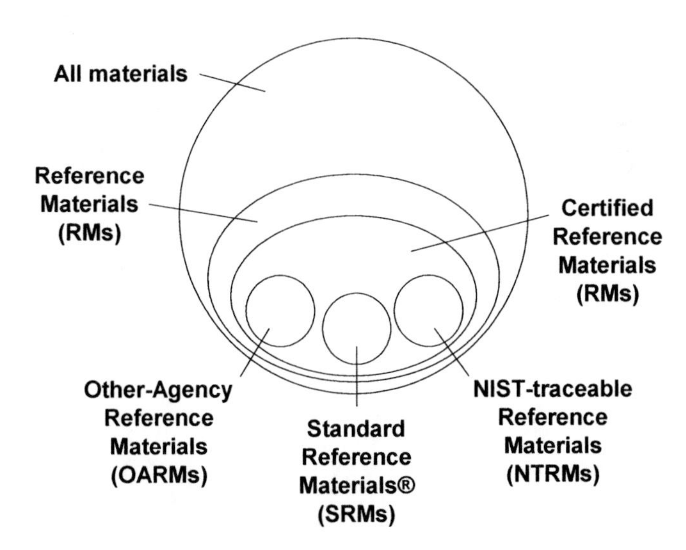

FIG. 1—*Relationships among all materials in our workplaces and environment and the subsets of those materials that have been designated as different categories of reference materials.*

To improve measurement science, NIST produces and distributes CRMs under the registered name *Standard Reference Material® (SRM)*. Thus, an SRM is a CRM that is issued by NIST. An SRM is certified for specific chemical or physical properties and is issued by NIST with a certificate that reports the results of the characterization and indicates the intended use of the material. The three technical categories of SRMs are chemical composition, physical properties, and engineering materials. The NIST Technology Services website (http://www.nist.gov/srm) describes the SRM program in detail.

NIST also prepares and characterizes CRMs under specific agreements with other agencies. These are known as *Other Agency Reference Materials (OARMs)*. Under such agreements, the entire lot of materials and the accompanying reports are transferred to the contracting agency. OARMs meet the ISO requirements for CRMs.

A *NIST Traceable Reference Material (NTRM)* is a CRM produced by a commercial supplier with a well defined traceability to the values of standards maintained by NIST. This traceability is established via criteria and protocols defined by NIST that are tailored to meet the needs of the metrological community to be served. The NTRM concept was established to allow NIST to efficiently respond to the increasing needs for high-quality reference materials. Reference material producers adhering to these requirements are allowed to use the NIST "NTRM" trademark to identify their products. NTRMs meet the ISO requirements for CRMs.

The first example of an NTRM was in the area of gas metrology. The gas NTRM program was established in 1992 in partnership with U.S. Environmental Protection Agency (EPA) and specialty gas companies as a means for providing end-users with the wide variety of certified gas standards needed to implement the "Emissions Trading" provision of the 1990 Clean Air Act. Gas NTRMs are produced and distributed by specialty gas companies with NIST oversight of the production and maintenance and direct involvement in the analysis. The gas standards prepared according to this program are related, within known limits of uncertainty, to specific gaseous primary standards maintained by NIST.

"Make-Your-Own" Reference Materials (MYOMs) are a new concept in CRMs that are made and evaluated (for the most part) outside the primary NIST laboratories for the specific purposes of the user. This type of reference material results from the complete blending of two other gravimetically aliquoted reference materials. Figure 2 illustrates the relationships between an existing reference material (or set of reference materials) and a MYOM. The certified or reference values for a MYOM are calculated using the corresponding values for the parent materials and the gravimetric dilution factor. To preserve the uncertainty levels of the parent materials in the MYOM, the uncertainty component associated with weighing and blending must be small compared to the uncertainties in the parent materials. MYOMs meet the ISO requirements for CRMs.

Types of Values and Modes of Certification

Each particular analyte, chemical or physical property, or quantity that is subject to measurement is referred to as a *measurand* [2]. For SRMs prepared by NIST, there are three levels of *values* that can be established for a particular measurand:

- A NIST Certified Value represents a value for which NIST has the highest confidence in its accuracy in that all known or suspected sources of bias have been fully investigated or accounted for by NIST. Certified values have associated uncertainties.

- A NIST Reference Value is a best estimate of the true value provided by NIST where all known or suspected sources of bias have not been fully investigated by NIST. Reference values have associated uncertainties.
- A NIST Information Value is a value that will be of interest and use to the SRM/RM user, but insufficient information is available to assess the uncertainty associated with the value. Information values do not have associated uncertainties.

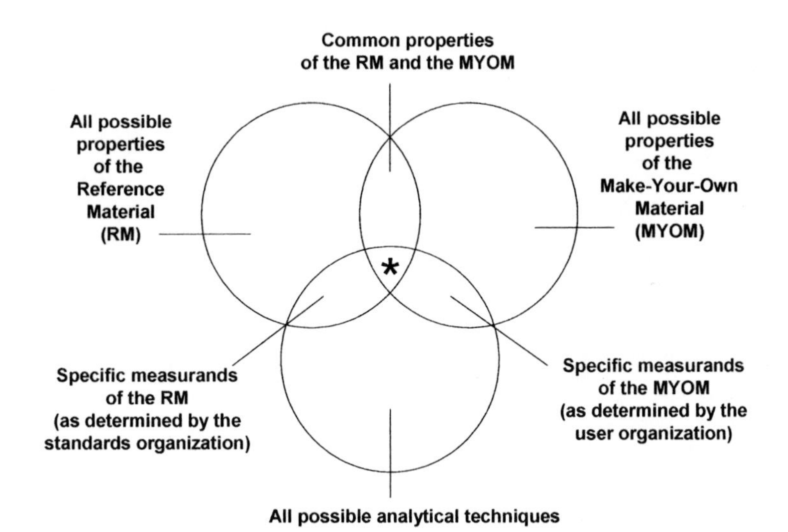

FIG. 2—*Relationships between an existing reference material (or set of reference materials) and a "Make-Your-Own" Reference Material (MYOM).*

The use and interpretation of certified values, reference values, and information values are related to an SRM user's requirements that his or her results be "*traceable to NIST.*" Although the phrase "traceable to NIST" is commonly used, it is strictly incorrect because results or values can only be traceable to stated reference results or values, not to institutes. Traceability is the property of the *result* of a measurement or the *value* of a standard whereby the result or value can be related to *stated references*, usually national or international standards, through an unbroken chain of comparisons all having stated uncertainties [2]. Thus, the phrase "traceable to NIST" can be thought of as "traceable to the results of measurements or the values of standards at NIST." Because the definition of traceability requires an unbroken chain of comparisons all having stated uncertainties, only certified values and reference values can be used as "stated references." Information values do not qualify as stated references, because they do not have uncertainties associated with them.

Table 1 summarizes the characteristics of seven *modes* by which certified values, reference values, and information values, or all three, can be developed [6]. The choice of mode(s) to be used in the value-assignment for any SRM for chemical measurements is based on previous experiences and knowledge of the specific matrix, analyte(s) of interest, current measurement capabilities, the quality of the analytical method's results, and the intended use of the material. The final designation of an assigned value for an SRM as a NIST certified value, NIST reference value, or NIST information value is based on the specific value-assignment mode used and the assessed quality of the resulting data relative to the intended use of the material.

TABLE 1—*Relationships between the seven modes of certification recognized by NIST and the type(s) of value (certified, reference, or information) that can be assigned by each certification mode [6].*

		Type of Value		
Mode	Description	Certified	Reference	Information
1.	Certification at NIST Using a Single Primary Method with Confirmation by Other Method(s)	•		
2.	Certification at NIST Using Two Independent Critically Evaluated Methods	•	•	
3.	Certification/Value-Assignment Using One Method at NIST and Different Methods by Outside Collaborating Laboratories	•	•	
4.	Value-Assignment Based On Measurements by Two or More Laboratories Using Different Methods in Collaboration with NIST		•	•
5.	Value-Assignment Based on a Method-Specific Protocol		•	•
6.	Value-Assignment Based on NIST Measurements Using a Single Method or Measurements by an Outside Collaborating Laboratory Using a Single Method		•	•
7.	Value-Assignment Based on Selected Data from Interlaboratory Studies		•	•

Certified Values

As summarized in Table 1, certified values result from any of Modes 1, 2, or 3, wherein either a *primary* method or two *independent* methods are used to determine an analyte.

- The concept of a *primary method* has been described by Moody and Epstein [7] as a "definitive method" and more recently articulated by the Consultative Committee for Amount of Substance (CCQM) as a method having the highest metrological properties, whose operation can be completely described and understood, for which a complete uncertainty statement can be written down in terms of units belonging to the International

System of Units (SI) [8]. In practice, a primary method has all of its potentially significant sources of error evaluated explicitly for the application of the method and the matrix under investigation [7]. NIST always combines primary methods with some means of confirmatory analysis. Such confirmation can be accomplished by the re-determination of certified constituents in other SRMs or CRMs of similar matrix and constituent concentration range, or by using a second method with appropriate controls. Confirmatory methods can be carried out either in NIST laboratories or by collaborating laboratories with appropriate experience.

- Methods are considered to be *independent* if they have completely different sources of error and variability. In practice this is rarely the case, but methods can usually be chosen so that the most significant sources of error are different. For example, where material dissolution might be a significant challenge to the analytical process, different sample preparation methods can be selected to minimize the chance that similar errors will be incurred. Instrumental methods can be considered to be independent if they rely on different physical, spectroscopic, or chemical phenomena to generate their respective analytical response. In all cases, the design of the certification plan ensures that all methods have the appropriate precision and accuracy for measurement of the target analyte(s) in the matrix.

Certification Mode 1 involves the use of a single primary method at NIST with confirmation by another method (or methods).

Certification Mode 2 involves the use of two or more critically evaluated *independent* NIST methods [9].

Certification Mode 3 is used when NIST does not have a suitable second independent method, and outside laboratories are selected to collaborate on the certification process. In such cases, NIST works very closely with the outside laboratory analysts to ensure that the details of the measurement protocol, data analysis, and reporting requirements are carried out according to NIST specifications.

Reference Values

Even though Modes 2 and 3 can result in a certified value, if the results of the methods do not agree sufficiently, NIST designates the assigned value as a reference value. Reference values can result from certification Modes 2–7.

Information Values

Certification Modes 4–7 can also result in a NIST information value. Such is the case when the agreement among data from multiple methods is not sufficient to estimate a reliable uncertainty. Information values provide users with supplemental information about the SRM composition. Certified values and reference values have corresponding uncertainties. Information values do not have corresponding uncertainties.

Application-Related Issues for Reference Materials

Reference Materials for Calibration

Reference materials used for the calibration of instruments or analytical methods are, in general, relatively simple mixtures of analytes whose concentrations are very well known. Often such reference materials are solutions, prepared from pure materials by highly accurate and precise means at high concentration. They are intended to be diluted either by volumetric or gravimetric procedures producing one or more working mixtures so that the resulting series of analyte concentrations spans the range of the intended analysis. Analyte concentrations in calibration reference materials are known to have such high accuracy and precision that they can be used to establish instrumental response calibration curves where the error can be assumed to reside predominantly in the instrumental readings. Minimizing uncertainty in the analyte concentration (the independent variable in the calibration procedure) avoids mathematical model complications known as the "error-in-X" case [10]. The assumption that error resides predominantly in the instrument reading (the dependent variable) is necessary, for instance, to estimate the slope and intercept of a straight line using simple linear regression.

Calibration reference materials can be the key to ensuring the quality of results for chemical analysis results provided that the chief source of error is in the calibration step of the analytical method. This is not often the case for the determination of beryllium in industrial materials or environmental samples. In these cases, the chemical matrix and physical form of the sample present far more significant sources of error due to incomplete chemical dissolution, interferences, and matrix effects. However, calibration reference materials can be effective tools in these cases when they are used in standard additions, internal standard, and matrix-matching schemes to correct for analytical bias.

Reference Materials for Method Validation

Reference materials intended to validate the entire analytical method are designed to represent, as faithfully as possible, the chemical matrix, physical form, and analyte levels of the target sample. By testing the validity of the entire analytical method, such reference materials ensure to a greater degree the quality of analytical results for real samples. For all their benefits, however, matrix reference materials are much more difficult to develop. Before one can begin to design the certification procedures for a matrix reference material, the process of selecting, processing, and accepting a candidate material must be addressed.

It is almost paradoxical that one must know a fair amount about real-world samples before the relevant matrix reference materials can be developed to qualitatively and quantitatively verify such knowledge. The target analytes in industrial and environmental samples can take on many forms, and one must understand and clearly state the physical and chemical form of the intended measurand. For example, free and bound forms of beryllium must be identified if a matrix reference material is to adequately challenge sample preparation steps. Some sort of screening analysis has to be used to characterize the chemical matrix of interest. In many cases, the sample collection method itself adds to the list of concomitant species that must be included in the design of the reference material. Industrial hygiene evaluations often employ air filters and wipe materials, which can interfere with either the sample preparation steps or can add to the list of species interfering with the detection of the analyte.

Some methods of analysis are intended to reproduce the action of natural environmental conditions. For instance rainwater and groundwater leaching effects can be mimicked by the extraction of analytes and interfering matrix species as an analysis preparation step. Not only do such applications require that reference materials and real samples be of a similar chemical matrix, but they must also be of similar physical form. Moreover, careful design work is necessary for reference materials to present the same challenges to the analytical chemistry process as the real samples. One straightforward way to ensure that a reference material matches all the critical parameters of the target sample types is to collect real material at a site of known contamination. A practical matter arises when the need to document the origin of a reference material conflicts with the site owner's desire to remain unidentified.

General Properties of Beryllium

Beryllium and its oxide are used in a wide variety of materials to take advantage of a number of its unique properties. As an alloying constituent, beryllium lends a high degree of dimensional stability and wear resistance to metals. Its superior transmittance to X-rays makes beryllium foil a good choice for spectrometer windows. Beryllium is also used in ceramics and in the nuclear industry for reactor moderation. Mining, extracting, and refining beryllium involves the production of fine powders, which potentially pose airborne and skin contact health risks. Evaluations of health effects, degrees of exposure, and the effectiveness of remediation measures depend on the accurate determination of beryllium using metrologically sound analytical chemistry methodologies.

Existing Beryllium Reference Materials

The information presented below on existing beryllium SRMs has been excerpted directly from the *NIST Certificate of Analysis* for each SRM described. The certificates of analysis routinely describe how the SRMs were intended to be used, how the SRMs were prepared, how uncertainties in the metrological properties were determined, the time period of the certifications for each of the materials, the manner in which NIST will maintain the certification, and instructions for use. If appropriate, an SRM is also accompanied by a *Material Safety Data Sheet (MSDS)* containing sections on material identification, hazardous ingredients, physical/chemical characteristics, fire and explosion hazard data, reactivity data, health hazard data (including emergency and first aid procedures), precautions for safe handling and use, and source data and other comments.

Beryllium Solution SRM

SRM 3105a is intended for use as a primary calibration standard for the quantitative determination of beryllium by analytical methods such as inductively coupled plasma spectrometry (both optical and mass) and atomic absorption spectrometry [11]. Each sales unit of SRM 3105a consists of five 10 mL sealed borosilicate glass ampoules each containing a 10 % nitric acid solution of beryllium prepared gravimetrically to contain a known mass fraction of beryllium. The current certified value of beryllium in SRM 3105a is 10.83 mg/g ± 0.07 mg/g [11]. That value is based on (1) gravimetric preparation and (2) inductively coupled plasma optical emission spectrometry (ICP-OES) using three independently prepared primary standards.

The uncertainty in the certified value is calculated as $U = (2\mu_c + B)$ mg/g where μ_c is the combined standard uncertainty calculated according to the ISO and NIST guidelines [12,13] and the procedure of Schiller and Eberhardt [14] for combining independent analytical methods. The value of μ_c is intended to represent, at the level of one standard deviation, the combined effect of uncertainty components associated with the gravimetric preparation and the analytical determinations. The quantity B is an allowance for between-method differences.

Note that the relative expanded uncertainty of the certified value for SRM 3105a is less than 0.7 %. This low level of uncertainty results from a combination of dissolution and gravimetric preparation using ultra-pure beryllium metal as well as the high-accuracy comparison of the candidate SRM solution with independently prepared NIST standard beryllium solutions. Besides the uncertainty components of these two methods of beryllium determination, the beryllium metal purity and its uncertainty were also taken into account. The use of two independent methods for the certification of beryllium in SRM 3105a represents one of the seven modes (Table 1) that NIST uses for value assignment of chemical reference materials.

Expiration of Certification—The current certificate of analysis for SRM 3105a states that the certification of SRM 3105a Lot No. 892707 is valid, within the measurement uncertainty specified, until 15 December 2006, provided the SRM is handled in accordance with the "Instructions for Use" given in the certificate. It is further stated that the certification is nullified if the SRM is damaged, contaminated, or modified.

Maintenance of Certification—The certificate describes how the certification will be maintained by NIST. NIST will monitor representative solutions from this SRM lot over the period of its certification. If substantive changes occur that affect the certification before the expiration of certification, NIST will notify the purchaser. Purchasers are advised to facilitate notification by returning the registration card that accompanies the certification.

Traceability to this SRM—The certificate states that calibration of analytical instruments or procedures for the determination of beryllium should be performed using standards that are traceable to this SRM. The traceability of standards to this SRM must be established through an unbroken chain of comparisons, each having stated uncertainties [2]. Comparisons are based on physical or chemical measurements proportional to the beryllium concentration. These may include various spectroscopic or classical methods of analysis. The gravimetric and volumetric dilution preparations are also considered to be methods of comparison. The uncertainties assigned to such traceable standards must include the uncertainty of this SRM appropriately combined with the uncertainties of all comparison measurements.

Instructions for Use Caution—The certificate cautions the user to handle the SRM in a safe manner (i.e., wear gloves and avoid accidental breakage or spillage during handling of acid solution SRMs that are contained in tip-sealed borosilicate glass ampoules with pre-scored stems).

Instructions for Preparation of Working Standard Solutions—The certificate describes how working standard solutions should be prepared by mass or by volume.

Blocks, Chips, and Rods of Beryllium-Copper Alloy

SRMs for beryllium in various solid copper alloys have been available from NIST since the early 1980s. High-purity copper was melted with several minor and trace elements to produce SRMs 1121, C1121, 1122, C1122, 1123, and C1123. The materials with the "C" designation were the chill-cast SRMs. Samples of the chill-cast material were 32 mm (1.25 in.) square by 19 mm (0.75 in.) thick. Wrought samples were 32 mm (1.25 in.) in diameter by 19 mm (0.75 in.) thick. Beryllium ranged from a certified value of 0.46 % in SRMs 1123 and C1123 to 1.92 % in C1121. Only SRM C1122, with beryllium at 1.75 %, is presently available for sale. The others in this series have been sold out and discontinued. This form of the alloy is intended primarily for calibration of X-ray and optical emission spectrometers.

Beryllium-copper alloys in chip form are intended for chemical analysis. SRMs 458, 459, and 460 are copper alloys with beryllium at 0.360 % in SRM 458, at 1.82 % in SRM 459, and at 1.86 % in SRM 460. They were prepared in cooperation with ASTM International and are in the form of chips sized between 0.50 mm and 1.18 mm sieve openings (35 mesh and 16 mesh). Certified values are also provided for the concentrations of aluminum, chromium, cobalt, iron, lead, magnesium, nickel, silicon, tin, and zinc in these SRMs. Cooperative analyses for certification were performed at the following laboratories: Armco Research and Technology, Armco, Inc., Middletown, OH; Brush Wellman, Inc., Elmore, OH; Colonial Metals Company, Columbia, PA; NGK Insulators Ltd., Handa City, Japan; NGK Metals Corp., Reading, PA; and Teledyne Wah Chang, Albany, OR. Information values are provided for the concentrations of antimony, copper, manganese, silver, sulfur, titanium, and zirconium.

Beryllium as a Trace Element

Beryllium is listed as a trace element in some standard NIST soil and sludge SRMs (e.g., SRM 1646a Estuarine Sediment, SRM 1944 New York/New Jersey Waterway Sediment, and SRMs 2586 and 2587 Trace Elements in Soil). Beryllium is assigned a reference value of 1.6 mg/kg with an expanded uncertainty of 0.3 mg/kg in SRM 1944. Concentrations of beryllium in the other SRMs are information values ranging from less than 1mg/kg to over 9 mg/kg.

Beryllium is also listed as information values in three coal fly ash materials (SRMs 2689, 2690, and 2691) as well as in SRM 1632c Bituminous Coal. The range of values is from about 1 mg/kg in the coal to 21 mg/kg in SRM 2689.

An information value of 5 μg/L is listed for beryllium in one of the series of toxic elements in urine SRMs (SRM 2670a).

Beryllium on Filter Media SRM (No Longer Available)

Beryllium on filter media was available at one time as SRM 2677a, but the certification of this SRM expired on 30 September 1999 and was not extended because the demand for these samples was low and the amount of laboratory effort to prepare the samples was high. SRM 2677a was intended primarily as an analytical standard for use in the determination of beryllium and arsenic in industrial atmospheres [15].

The filters were of the mixed cellulose ester type, and were 37 mm in diameter with a pore size of 0.8 μm. Each filter was prepared by depositing a 50 μL aliquot of an appropriate composite solution of Be and As onto the filter, followed by drying. The composite solutions were prepared gravimetrically by mixing together appropriate amounts of a standard beryllium solution (prepared from high-purity Be metal) and a standard arsenic solution (prepared from SRM 83d, As_2O_3). In the preparation of the arsenic standard, As^{+3} was oxidized to As^{+5} with bromine and was expected to be present on the filters as the arsenate. The blank filters were prepared by adding a 50 μL aliquot of the dilute mixed acid (HNO_3 and H_2SO_4) solution to each filter.

SRM 2677a consisted of a set of ten membrane filters (two at each concentration level), packaged in five Petri dishes, with each Petri dish containing two (i.e., duplicate) filters from one of the following five ranges of concentrations (in micrograms per filter): Level I (0.129 ± 0.003 Be and 0.269 ± 0.006 As), Level II (0.643 ± 0.015 Be and 2.69 ± 0.065 As), Level III (2.58 ± 0.06 Be and 26.92 ± 0.65 As), Level IV (0.050 ± 0.001 Be and 0.101 ± 0.002 As), and Blank (≤ 0.0005 Be and ≤ 0.0005 As).

The certified values for SRM 2677a were based on gravimetric measurements made during the production of four stock solutions used to impregnate the filters and on measurements of the amount of stock solution deposited on the filters. The listed uncertainties were expressed as two standard deviations for a single filter, and included the uncertainties of the stock solutions used in the preparation of the filters.

The certificate for SRM 2677a noted that, in all instances, an entire filter must be dissolved for each set of measurements because the metals may not be uniformly distributed on the filter.

Insights and Examples from Non-Beryllium Reference Materials

Understanding the preparation methods used over the years by NIST for soils, dusts, and other materials gives an indication of the range of strategies that might be used to prepare new beryllium SRMs. There are both specific and general lessons to be learned. A general lesson is related to ensuring SRM homogeneity. For example, reference material preparation methods should be designed to ensure homogeneity of the material at whatever minimum aliquot size is required. For bulk materials like soils and sediments, the minimum sample size can be in the hundreds of milligrams range. However, when sampling schemes for real samples involve the wiping of surfaces, sample sizes can be in the microgram range, presenting a very difficult challenge to ensuring homogeneity of the reference material. Examples of historically important and relevant NIST SRMs are given below.

An Urban Particulate Matter SRM

Originally certified in 1978, SRM 1648 Urban Particulate Matter still sells at a rate of over 100 units (2 g of material per bottle) per year. This SRM was prepared from airborne ambient dust collected from the St. Louis area using a bag house specifically designed for the purpose. Collection took place over a period in excess of twelve months. The material was removed from the filter bags, combined into a single lot, screened through a fine-mesh sieve to remove extraneous materials, and thoroughly blended in a V-blender. In this case, the collected material was already in the same physical form as the intended sample.

A Lead Contamination Indoor Dust SRM

When the candidate reference material is mixed with other undesirable material, more extensive physical preparation steps are necessary. SRM 2584 is a dust material that was collected from interior living spaces. Approximately 65 % of the material was obtained from households involved in lead poisoning intervention programs in which vacuum cleaners with high efficiency air filters were used to remove dust and other surface debris from homes where cases of lead poisoning had occurred. This material was mixed with low-lead-level material taken from conventional vacuum cleaner bags from households not identified as having a lead contamination problem. All dust bags and their contents were radiation-sterilized. The material from each bag was then mixed and tumbled in a modified food processor using chopping blades and a compressed air jet. While still tumbling, the dust was separated from unwanted debris by vacuuming through a series of screens into a clean HEPA vacuum cleaner. The dust collected in this manner was then screened through a 90 μm stainless steel sieve using vibration and a vacuum. Processed sub-lots of approximately 5 kg each were set aside and analyzed for lead by X-ray fluorescence in order to develop a blending protocol for the target lead concentration. Selected high- and low-level sub-lots were blended in a cone blender and then bottled.

Soils and Sediments SRMs

For candidate materials that are collected in bulk, a change in the physical form and particle size distribution, or both, is often necessary. For soils and sediments, extensive grinding, milling, and sieving may be required to prepare the material for blending. Effective homogenization requires a narrow particle size distribution; however, the material becomes less useful for bulk physical properties performance measures. For natural-matrix reference materials, the emphasis at NIST is usually on chemical composition, so most materials undergo extensive treatments to obtain homogeneous samples with small mean particle sizes.

For example, NIST and the U.S. Geological Survey (USGS) collaborated in 1992 to produce a series of three SRM soil materials: SRM 2709 San Joaquin Soil, SRM 2710 Montana Soil with highly elevated trace element concentrations, and SRM 2711 Montana Soil with moderately elevated trace element concentrations. Each of these materials is certified for over 25 elements, with information values for more than 30 elements and leachable concentrations using U.S. EPA Method 3050 for flame atomic absorption spectrometry (FAAS) and inductively coupled plasma optical emission spectrometry (ICP-OES). To ensure homogeneity at the 250 mg sample size, the material for these SRMs was subjected to a series of preparation steps, including gross physical separation from debris and pre-drying in an air oven and for three days at room temperature. The material was then passed over a vibrating 2 mm screen to remove plant material, rocks, and large chunks of aggregated soil. Material remaining on the screen was disaggregated and rescreened. The combined material passing the screen was ground in a ball mill to pass a 74 μm screen and blended for 24 h.

Biological Matrix SRMs

The concentrations of key elements in biological and botanical matrix SRMs are often so low that special techniques are needed to reduce the size of particles and narrow their size distribution while avoiding contamination from the preparation equipment. SRM 1566b Oyster Tissue is an example of such a material. This material was initially freeze-dried, broken into small pieces, and blended in a mixer with titanium blades. The final step took place in a jet mill specially designed for this purpose. The sample was entrained in two high-speed streams directed at each other so that the sample collided with itself, fracturing the particles without abrasion from foreign material.

Air Sampling or Wipe Sampling SRMs

Many environmental and industrial hygiene sampling protocols specify sampling using air filters or wipes. Over the years, NIST has investigated ways to deposit candidate material on filters and wipes with some success. SRM 2783 Air Particulate on Filter Media and SRM 2679a Quartz on Filter media were both prepared by suspending homogeneous particulate material in a liquid, depositing a measured amount on each filter blank, and carefully drying under clean conditions. The process is quite tedious, and the loaded filter must be handled very carefully. While binders could be used to improve the physical stability of these materials, they can introduce chemical matrix effects not encountered in real samples.

Steps for New Beryllium Reference Materials

Figure 3 illustrates the total life cycle process for reference materials. Necessary actions for preparing new reference materials include consensus on the desired chemical and physical forms of beryllium; market analyses of demand for the materials; identification of candidate industrial or laboratory-produced samples of the materials; selection of samples that meet criteria for uniformity, physical form, measured quantities, and continued availability; development of production protocols for collection and preparation of the materials, including adequate provisions for occupational health and environmental protection; incorporation of these materials into a supply, distribution, and cost-recovery infrastructure; and continued feedback and information sharing to ensure that the reference materials are meeting user needs or are modified as necessary. Issues for applying each of these steps to beryllium are discussed below.

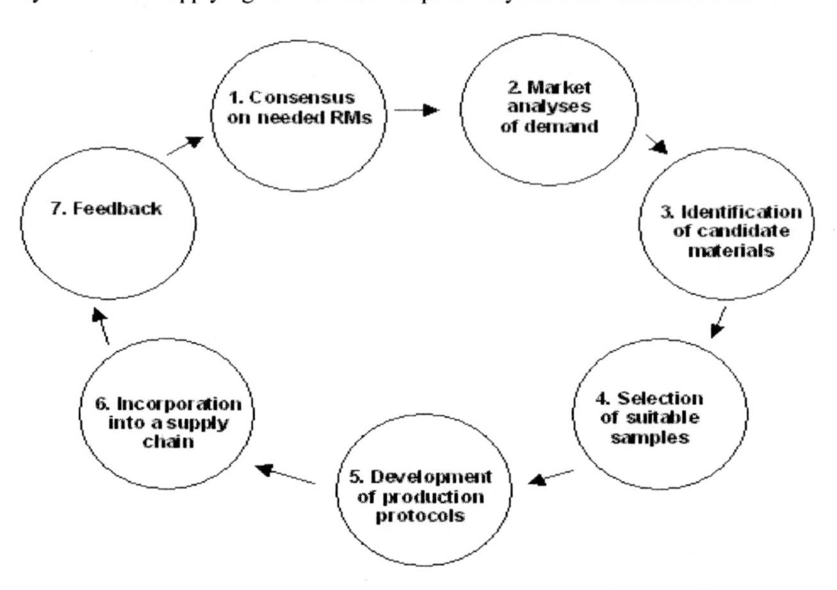

FIG. 3—*Overview of the integrated life cycle process for reference materials.*

Step 1: Consensus on Candidate Beryllium Reference Materials

As noted above, current beryllium SRMs are limited to standard spectroscopy solutions; blocks, chips, and rods of copper-beryllium alloy; and beryllium as a trace element in some standard soil-sludge, coal fly ash, and urine reference materials. Beryllium on filter media was available at one time, but is not currently available. It would be useful to have SRMs for beryllium-containing ores, beryllium oxide, beryllium metals, beryllium-containing alloys, and beryllium in biological samples. Continued dialogues involving the beryllium industry, regulatory, and research stakeholder communities are needed to develop and prioritize a complete list of candidate reference materials.

Step 2: Market Analyses of Demand

Decisions on which beryllium materials should be developed into reference materials first (or ever) are likely to be driven by two types of market analyses of demand for the materials, in combination with feasibility considerations for how, where, and by whom the reference materials will be made.

The first analysis involves simple economics: How quickly will projected revenues meet or exceed projected costs? In its normal decision making, NIST uses a cost-recovery formula by evenly dividing the costs of production over the number of SRM units that are expected to sell in five years. If the economics are favorable, the decision to proceed would simply require the expertise and input to complete all remaining steps of the reference material lifecycle process.

The second analysis disregards simple economics and focuses on what materials are needed to answer critical health and safety questions. Costs of producing and distributing the reference materials would most likely have to be underwritten by sponsoring agencies. In such cases, feasibility issues (how, where, by whom) would still be the final deciding factors.

Step 3: Identification of Candidate Materials

Identification of candidate industrial or laboratory-produced samples of beryllium materials has recently focused on two materials: well characterized powders of product type I-400 beryllium metal, which includes particles in the respirable size range, and well characterized powders of product type UOX-125 beryllium oxide, which consists of aggregates of 200 nm diameter primary particles. Additional information about these two candidate materials can be found in recent publications by Stefaniak et al. [16,17,18]. Additional discussion and research will be required to identify candidate materials of beryllium-containing ores; beryllium oxide in the form of larger, compact particles; beryllium-containing alloys, including copper, nickel, aluminum, and other materials; and beryllium in biological samples, including lung, skin, and other organs and tissues.

Step 4: Selection of Suitable Samples

Selection of samples that meet acceptable criteria for uniformity, physical form, measured quantities, and continued availability would be straightforward for the proposed beryllium metal and beryllium oxide materials. Type I-400 beryllium metal and type UOX-125 beryllium oxide are industrial products, have defined production pedigrees, have been extensively characterized (as noted above), and can be obtained in large quantities (i.e., more than tens of kilograms).

They can be size-separated in the laboratory using well established and published techniques [19].

Step 5: Development of the Production Protocol to Establish and Document the Pedigree of the Reference Material

Production protocols would specify precisely how the material would be prepared so that the pedigree of the material is appropriately established and documented. This would include the specification of the required measurands and the procedures for determining the measurands. Development of production protocols for collection, preparation, and characterization of the new beryllium reference materials would involve not only aerosol and materials science considerations, but also substantial attention to adequate provisions for occupational health and environmental protection. Original preparation of size-selected samples of the I-400 and UOX-125 beryllium materials was done in the 1980s and 1990s in projects sponsored by the U.S. Department of Energy (DOE) in the specialized and highly controlled inhalation toxicology research laboratories at Lovelace Respiratory Research Institute (LRRI) in Albuquerque, New Mexico. The beryllium aerosol facilities at LRRI have been decommissioned and the facilities have been turned to other uses. NIST has some capabilities for working with toxic materials, but does not currently have the capacity to undertake production of beryllium powder SRMs. Interagency discussions of how and where beryllium reference materials could be produced are underway among representatives of interested federal agencies including NIST, NIOSH, and DOE. The production options for new beryllium CRMs include the categories of Standard Reference Materials, NIST-traceable reference materials, or Other-Agency Reference Materials.

Once a candidate material has been accepted and its homogeneity assessed, the design of the certification program can be determined, keeping in mind the intended use of the material and the capabilities of the analytical methods. If the material were made and evaluated (for the most part) outside the primary NIST laboratories, it might have the attributes of a "Make-Your-Own" reference material as described in Fig. 2. For example, as is done currently, the specific surface area of the new beryllium reference material would be determined by comparison to existing NIST surface area SRMs (none of which are beryllium); the density of the beryllium reference material would be determined in comparison to non-beryllium reference materials; and the particle physical size would be determined from microscopy that is calibrated with non-beryllium materials. Properties such as crystalline structure would be determined by comparison to known beryllium references.

Note that, as with all materials, many of the measurands for beryllium reference materials can be experimentally determined by more than one technique. For example, beryllium particle solubility can be assessed in simulated lung fluid or in simulated phagolysosomal fluid, as well as by observation of dissolution behavior in beryllium-exposed laboratory animals. Similarly, particle "size" can be assessed by cascade impaction (aerodynamic diameter), by diffusion (thermodynamic diameter), by light-scattering, or by electrical mobility. These techniques are based on fundamentally different processes, requiring differing interpretation or leading to different results. When such information is communicated, in addition to any normally reported degree of confidence or data distribution that may be quoted, it will be critical to state the means by which the information was derived or determined.

Step 6: Incorporation into a Supply Chain

Incorporation of new beryllium reference materials into a supply, distribution, and cost-recovery infrastructure will require designation of a responsible organization or network of organizations. Preparation, packaging, and shipping could be done by an organization different from the organization responsible for listing the available materials, taking orders, and conducting billing.

Step 7: Feedback

Continued feedback and information sharing is critical to ensuring that the reference materials are meeting user needs or are modified as necessary. It is likely that applications of any new reference materials will contribute to a better understanding of the needs for other reference materials. In addition, the results of laboratory studies and field applications of the reference materials may increase the number and quality of the certified, reference, and information values for measurands of interest for these materials.

Conclusion

A number of opportunities exist for identification and development of new beryllium reference materials. Taking advantage of these opportunities will require multi-disciplinary and multi-organizational collaboration. Past experiences can be built upon and new relationships and capabilities can be conceived and implemented. Beryllium is not the only toxic material for which a broader spectrum of reference materials would be useful. Lessons from new initiatives in beryllium can inform our strategies for dealing with other toxic agents over a broad spectrum of organizational and disciplinary lines.

Acknowledgments

The authors gratefully acknowledge the many individuals presently or formerly at NIST who participated in the preparation, statistical assessment, certification, and issuance of the many SRMs described in this report, including T. A. Butler, T. C. Rains, T. A. Rush, L. L. Yu, L. J. Wood, and B. S. MacDonald.

The findings and conclusions in this report are those of the authors and do not necessarily represent the views of the National Institute of Standards and Technology or the National Institute for Occupational Safety and Health.

References

[1] ISO Guide 30, Terms and Definitions Used in Connection with Reference Materials, International Organization for Standardization, Geneva, Switzerland, 1992.
[2] International Vocabulary of Basic and General Terms in Metrology, International Organization for Standardization, Geneva, Switzerland, 1993.
[3] ISO Guide 34, General Requirements for the Competence of Reference Material Producers, International Organization for Standardization, Geneva, Switzerland, 2nd ed., 2000.
[4] ISO Guide 35, Certification of Reference Materials—General and Statistical Principles, International Organization for Standardization, Geneva, Switzerland, 2nd ed., 1989.

[5] ISO Guide 31, Reference Materials—Contents of Certificates and Labels, International Organization for Standardization, Geneva, Switzerland, 2nd ed., 2000.
[6] May, W. E., et al., "Definitions of Terms and Modes Used at NIST for Value-Assignment of Reference Materials for Chemical Measurements," *NIST Special Publication SP 260-136*, 2000.
[7] Moody, J. R. and Epstein, M. S., "Definitive Measurement Methods," *Spectrochim. Acta,* Vol. 46B, 1991, pp. 1571–1575.
[8] Minutes from the Fifth Meeting (February 1998) of the Comité Consultatif pour la Quantité de Matière (CCQM) of the Bureau International des Poids et Mesures (BIPM), Sevres, France, 1998.
[9] Epstein, M. S., "The Independent Method Concept for Certifying Chemical-Composition Reference Material," *Spectrochim. Acta*, Vol. 46B, 1991, pp. 1583–1591.
[10] Fuller, W. A., *Measurement Error Models,* John Wiley, New York, 1987.
[11] Certificate of Analysis - Standard Reference Material® 3105a - Beryllium Standard Solution - Lot No. 892707, National Institute of Standards and Technology, Gaithersburg, MD, originally issued on 08 December 1998, extension of the certification period on 11 January 2002, and further extension of the certification period on 17 November 2003.
[12] Guide to the Expression of Uncertainty, International Organization for Standardization, Geneva, Switzerland, 1st ed., 1993.
[13] Taylor, B. N. and Kuyatt, C. E., "Guidelines for Evaluating and Expressing the Uncertainty of NIST Measurement Results," NIST Technical Note 1297, U.S. Government Printing Office, Washington, DC, 1994; available at http://physics.nist.gov/Pubs/.
[14] Schiller, S. B. and Eberhardt, K. R.; "Combining Data from Independent Chemical Analysis Methods," *Spectrochimica Acta,* Vol. 46B, 1991, pp. 1607–1613.
[15] Certificate of Analysis - Standard Reference Material® 2677a - Beryllium and Arsenic on Filter Media, National Institute of Standards and Technology, Gaithersburg, MD, originally issued on 01 May 1990, editorial change on 08 February 1994, and addition of an expiration date on 03 December 1998. Available online at https://srmors.nist.gov/certificates/view_cert2gif.cfm?certificate=2677a.
[16] Stefaniak, A. B., et al., "Surface Area of Respirable Beryllium Metal, Oxide, and Copper Alloy Aerosols and Implications for Assessment of Exposure Risk of Chronic Beryllium Disease," *Am. Ind. Hyg. Assoc. J.*, Vol. 64, 2003, pp. 297–305.
[17] Stefaniak, A. B., et al., "Characterization of Physicochemical Properties of Beryllium Aerosols Associated with Prevalence of Chronic Beryllium Disease," *Journal of Environmental Monitoring,* Vol. 6, 2004, pp. 523–532, DOI:10.1039/b316256g.
[18] Stefaniak, A. B., et al., "Characteristics of Beryllium Oxide and Beryllium Metal Powders for use as Reference Materials," *Journal of ASTM International,* Vol. 2, No. 10, 2005.
[19] Hoover, M. D., Castorina, B. T, Finch, G. L, and Rothenberg, S. J., "Determination of the Oxide Layer Thickness on Beryllium Metal Particles," *Am. Ind. Hyg. Assoc. J.*, Vol. 50, 1989, pp. 550–553.

Journal of ASTM International, November/December 2005, Vol. 2, No. 10
Paper ID JAI13174
Available online at www.astm.org

A. B. Stefaniak,[1] *M. D. Hoover,*[1] *G. A. Day,*[1] *A. A. Ekechukwu,*[2] *G. E. Whitney,*[3] *C. A. Brink,*[3] *and R. C. Scripsick*[3]

Characteristics of Beryllium Oxide and Beryllium Metal Powders for Use as Reference Materials

---

ABSTRACT: Laboratory evaluations of commercially available powders of beryllium oxide (BeO) and beryllium metal, with special emphasis on type UOX-125 BeO and type I-400 beryllium metal, are ongoing to develop reference materials for evaluating analytical chemistry digestion methods and facilitating occupational health studies. Measured properties of the powders include morphology, size, density, specific surface area (SSA); crystalline and elemental composition; surface chemistry, and *in vitro* dissolution in hydrochloric acid (HCl) (pH 1), phagolysosomal simulant fluid (PSF) (pH 4.5), and serum ultrafiltrate (pH 7.3). The powders were also used to evaluate the digestion and recovery efficiencies for commonly used U.S. Environmental Protection Agency (EPA), National Institute for Occupational Safety and Health (NIOSH), and Occupational Safety and Health Administration (OSHA) standard analytical methods. UOX-125 BeO powder has high-purity and aggregate cluster morphology with SSA independent of aerodynamic particle cluster size, which results in dissolution kinetics that are independent of cluster size. I-400 beryllium metal powder has high-purity and compact particle morphology with SSA that increases as particle size decreases, which causes size-dependent dissolution kinetics (i.e., smaller particles dissolve more quickly than larger particles). The PSF and HCl chemical dissolution rate constants ($g \cdot cm^{-2} \cdot day^{-1}$) for the BeO powder were a factor of 10 lower than for the metal powder. Concomitantly, the EPA and NIOSH analytical methods, which used aggressive digestion procedures (e.g., microwave-assisted sample digestion or perchloric acid), gave more complete recovery of beryllium from BeO compared to the OSHA analytical method. Our characterization data suggest that these BeO and metal powders hold promise for use as analytical reference materials. We recommend continued laboratory collaborations to evaluate and apply these BeO and beryllium metal powders as analytical reference materials.

KEYWORDS: beryllium oxide, beryllium metal, reference material, digestion, particle

Introduction

Exposure to beryllium particles is associated with development of chronic beryllium disease (CBD) [1–4], a progressive lung disease characterized by non-caseating granulomas and fibrosis that occurs in individuals who are sensitized to beryllium [5]. The National Institute for Occupational Safety and Health (NIOSH) has estimated that as many or more than 130 000 workers in the United States are potentially exposed to beryllium under a wide variety of circumstances [6], making determination of beryllium levels in the environment of significant interest to the public health community.

Methods for quantifying levels of elements in environmental samples can be divided into two sequential steps: sample preparation (digestion) and sample analysis (determination of elemental mass). During sample preparation, an environmental sample and its associated matrix are digested to ensure the analyte is completely dissolved prior to analysis. For particulate samples,

---

Manuscript received 2 March 2005; accepted for publication 10 June 2005; published November 2005. Presented at ASTM Symposium on Beryllium: Sampling and Analysis on 21-22 April 2005 in Reno, NV.
[1] National Institute for Occupational Safety and Health, Morgantown, WV 26505.
[2] Savannah River National Laboratory, Aiken, SC 29808.
[3] Los Alamos National Laboratory, Los Alamos, NM 87545.

if a fraction of the analyte remains in undissolved particulate form after digestion, underestimation of mass will occur during analysis. If a digestion procedure is not sufficient to completely solubilize all forms and sizes of particles in a sample, the analytical method will fail to measure the elemental mass contribution of the incompletely dissolved particles. In the case of poorly soluble metals, complete digestion of all particulate to their dissolved form during sample preparation is therefore necessary to yield accurate and precise determinations of elemental mass levels in samples.

Quantitative analytical methods for beryllium are validated using standard reference materials containing beryllium. Existing beryllium standard reference materials are prepared using beryllium acetate, a soluble and easily digested compound that is either analyzed directly or applied to a sampling substrate such as a paper or membrane cellulose filter. In contrast, particulate beryllium in the form of beryllium oxide (BeO) and metal is poorly soluble [7–9] and can therefore be more difficult to digest completely. This difference in dissolvability is not assessed by the existing soluble beryllium standard reference material. As illustrated in the work reported here, recovery of beryllium from samples that contain particulate BeO and metal may not be fully known for methods that rely on the existing, soluble beryllium standard reference material to assess method recovery. Therefore, particulate beryllium reference materials are needed to supplement validation of a beryllium analytical method.

Currently, there is no standard reference material of high-purity particulate beryllium in any chemical form. Achieving and maintaining proficiency for analysis of all forms of beryllium material will require a suite of beryllium standard reference materials ranging from solutions to insoluble particulate forms. The existing beryllium acetate standard reference material could be used to evaluate instrument analysis accuracy (without concerns for digestion efficiency errors), and new particulate beryllium standard reference materials could be used to evaluate digestion and instrument analysis accuracy. The availability of BeO and metal powder reference materials in a range of particle sizes representative of what is encountered in the workplace, as powder or suspended in an appropriate matrix, would also permit industrial hygienists and chemists to assess the accuracy of beryllium analytical method and laboratory procedures by preparing and submitting spike samples blind with their environmental samples. Well-characterized and readily available reference materials of beryllium powders, having a range of particle sizes of concern for adverse health effects, could also be used to study the influences of beryllium physicochemical form and to eliminate material properties as a source of variability in inhalation toxicology, dermal exposure investigations, molecular biology, and immunology studies.

The purpose of this paper is to report our efforts to improve the scientific basis for quantification of beryllium in environmental samples and beryllium occupational health studies by conducting laboratory evaluations of commercially available powders of BeO and beryllium metal as reference materials, with special emphasis on type UOX-125 BeO and type I-400 beryllium metal powders (Brush Wellman Inc., Elmore, OH) that have been aerodynamically size-separated in the laboratory. Results from our characterization of these bulk powders, and aerodynamically size-separated subsets of these powders, suggest that these BeO and metal powders hold promise for use as analytical reference materials.

Materials and Methods

Size Separation of Beryllium Powders

Bulk samples of BeO (product type UOX-125) and beryllium metal (product type I-400) powder were obtained from Brush Wellman Inc. (Elmore, OH). Note that numerical designations refer to mesh sizes of screens through which powders were sieved by the manufacturer. These powders were chosen for study because they are primary feed materials for manufacturing beryllium metal and oxide ceramic parts, and CBD has been found in workers who form or machine these parts [10–12].

Details of the aerosol generation and aerodynamic size separation procedure for the powders have been described [13–15]. Briefly, bulk powders were aerosolized using a dry powder blower (Model 175, DeVilbiss, Somerset, PA) and the aerosol aerodynamically size-separated using a 5-stage aerosol cyclone [16] operated at 24 L min^{-1} and 20°C, followed by an electrostatic precipitator (ESP) (Mine Safety Appliances, Pittsburgh, PA). The aerodynamic cutoff diameters for the 5-stage aerosol cyclone and ESP used to size-separate the beryllium aerosols were >6, 2.5, 1.7, 0.9, 0.4, and ≤0.4 µm for stages 1 to 5 and the ESP, respectively.

Particle Physicochemical Characterization Techniques

A suite of analytical techniques was used to characterize physicochemical properties of the BeO and metal powders [9,14,15] (Table 1). Transmission electron microscopy (TEM) (Model CM30, Philips Electron Optics, Eindhoven, Netherlands) was used to assess particle morphology and size from samples prepared on 300-mesh copper grids coated with a lacey carbon substrate (Ted Pella Inc., Redding, CA). Nitrogen gas adsorption (Monosorb Model MS-16 Automated Direct-Reading Surface Area Analyzer, Quantachrome Corp., Syossett, NY) was used to determine powder specific surface area (SSA).

TABLE 1—*Beryllium powder characterization techniques.*

Technique[A]	Objective	Comments	Mass[B]
TEM	Morphology, size	Properties of individual or multiple particles	pg
Gas adsorption	Surface area	Total surface area of particle sample by gas adsorption	mg
XRD	Crystalline composition	Bulk analysis of constituents at 1 % or more by weight	mg
TEM-SAD	Crystalline composition	Properties from a selected viewing area; typically multiple particles	ng
TEM-µD	Crystalline composition	Properties of individual particles	pg
TEM-EDS	Elemental composition	Elements from C to U in a selected viewing area; typically multiple particles	ng
TEM-EELS	Elemental composition	Properties of individual particles	pg
XPS	Elemental composition	Surface analysis of chemical composition	mg
Pycnometry	Density	Density size by gas adsorption or gradient ultracentrifugation	mg
NAA	Oxide surface layer	Estimated thickness of the BeO surface layer on metal particles	mg
In vitro	Solubility	Quantification of chemical dissolution rate constant	mg

[A] TEM = Transmission electron microscopy.
XRD = X-ray diffraction.
TEM-SAD = Transmission electron microscopy-selected area electron diffraction.
TEM-µD = Transmission electron microscopy-micro electron diffraction.
TEM-EDS = Transmission electron microscopy-energy dispersive spectrometry.
TEM-EELS = Transmission electron microscopy-electron energy loss spectrometry.
XPS = X-ray photoelectron spectroscopy.
NAA = Neutron activation analysis.
[B] Mass indicates approximate amount required for an individual analysis.

X-ray diffraction (XRD) (Model XDS2000 powder diffractometer, Scintag, Inc., Sunnyvale, CA), TEM-selected area electron diffraction (TEM-SAD) (Philips Electron Optics), and TEM-microelectron diffraction (TEM-μD) (Philips Electron Optics) were used to qualitatively identity crystalline chemical constituents of samples. Note that XRD and TEM-SAD were used to determine constituents on powder samples and subsets of powder samples, whereas TEM-μD was used to determine constituents of individual particles. TEM-SAD and TEM-μD analyses were performed using the same grid samples prepared for TEM morphology and size analyses.

TEM-energy dispersive x-ray spectrometry (TEM-EDS) (germanium detector, Princeton Gamma-Tech, Princeton, NJ), TEM-electron energy loss spectrometry (TEM-EELS) (Model 766 DigiPEELS, Gatan, Pleasanton, CA), and x-ray photoelectron spectroscopy (XPS) (Model PHI 5600, Perkin-Elmer Corp., Eden Prairie, MN) were used to qualitatively identify elemental constituents of powder samples. TEM-EDS and TEM-EELS analyses were performed using the same grid samples prepared for the previously described TEM analyses. Note that elemental beryllium (atomic number 4) could not be detected using our TEM-EDS system, but elements in the sample having an atomic number greater than carbon (atomic number 6) could be identified. XPS is a surface technique that was used to determine the chemical composition and relative percent abundance of elements on the outer 50 to 75 Å-thick surface layer of powder sample subsets. The presence and estimated thickness of an oxide layer on the surface of I-400 beryllium metal particles, as a function of particle size, was previously evaluated by a combination of density measurements (gas pycnometry and gradient ultracentrifugation), SSA determinations, and oxygen-content determination by neutron activation analysis (NAA) [13].

Values of the chemical dissolution rate constant (k) for the powders were assessed *in vitro* using a static dissolution technique [17]. Rates of BeO and metal powder dissolution were previously determined in a range of solvents at 37°C, including 0.1 N hydrochloric acid (pH 1), phagolysosomal simulant fluid (pH 4.5), and serum ultrafiltrate (pH 7.3) [7–9].

Mercer Dissolution Theory

The dissolution theory of Mercer [18] was used to evaluate the comparative solubility and particle dissolution lifetimes of the BeO and metal powders. From Mercer's dissolution theory, the initial fractional dissolution rate at which a single particle suspended in a liquid medium dissolves is:

$$\left.\frac{df}{dt}\right]_{t=0} = -k\frac{6}{\rho D_0} \tag{1}$$

where, k = chemical dissolution rate constant $\left[\frac{\text{mass}}{\text{area} \cdot \text{time}}\right]$

ρ = particle density

D_0 = initial particle diameter

It follows that a particle is completely dissolved when:

$$t = \frac{3\alpha_v \rho D_0}{\alpha_s k} \tag{2}$$

where, t = time

α_v = volume shape factor, *e.g.* $\frac{\pi}{6}$ for a sphere

α_s = surface shape factor, *e.g.* π for a sphere

Evaluation of Digestion and Recovery Efficiencies of Commonly Used Analytical Methods

The BeO and metal powders were used to evaluate the recovery of beryllium digested by several standard analytical methods:

- U.S. Environmental Protection Agency (EPA) SW-846 Method 3051: Microwave assisted acid digestion of sediments, sludges, soils and oils [19],
- U.S. Occupational Safety and Health Administration (OSHA) method 125G: Metal and metalloid particulates in workplace atmospheres [20], and
- NIOSH method 7300: Elements by ICP [21].

In addition, for comparison, an aqueous beryllium standard reference material was used to evaluate the recovery of beryllium digested by:

- EPA SW-846 Method 3015: Microwave assisted acid digestion of aqueous samples and extracts [19].

Note that all spiked samples used to evaluate the recovery of beryllium from powders digested by these standard analytical methods contained mass levels that, at a minimum, exceeded the respective method reporting limit for beryllium by a factor of two. Spike sample matrices were varied among analytical laboratories, permitting challenge with several different matrices that were representative of those commonly used during environmental monitoring.

Modified EPA Method 3051—Recovery of beryllium from BeO and metal powder digested by a modified EPA Method 3051 was performed by a commercial laboratory. Suspensions of known concentration were prepared by adding phosphate buffered saline (PBS) solution to known masses of powder in glass scintillation vials and subjecting the vials to ultrasonic sonication for 15 min. Each suspension was occasionally shaken by hand to break apart agglomerates and was shaken vigorously immediately prior to pipetting a known amount of suspension (50–850 μg BeO collected in stage 2, 3, or 4 of the aerosol cyclone, 60–2300 μg metal powder collected in stage 2, 3, or 4 of the aerosol cyclone) onto 37-mm diameter cellulose filter support pads (stock number AP10, Millipore, Bedford, MA) [8]. Each aliquot used to prepare a spike sample was drawn from the center of the suspension, and any liquid remaining on the pipette tip was removed with a lint-free wipe prior to dispensing the material. Spike samples were submitted blind to the laboratory for quantification.

Spiked filter samples were generally prepared in replicates of 3–5 at each mass level. Each spiked sample (n = 49 BeO, n = 34 metal powder) was placed in a Teflon microwave digestion vessel. Ten mL of concentrated nitric acid and 2.5 mL of concentrated hydrochloric acid were added to each sample. The vessels were sealed, heated in a microwave for 30 min, cooled, transferred to 50-mL centrifuge cones, and diluted to 50 mL with ultrapure water (ASTM Type I). All samples were analyzed by inductively coupled plasma-mass spectrometry (ICP-MS) (Elan 6000, Perkin Elmer, Wellesley, MA).

Modified OSHA Method 125G—Recovery of beryllium from BeO powder digested by a modified OSHA Method 125G was performed by a private laboratory. A suspension of known concentration was prepared by adding PBS to BeO powder in a glass scintillation vial and subjecting the vial to ultrasonic sonication for 15 min. The suspension was shaken vigorously by hand immediately prior to pipetting a known amount of suspension (0.05–10.0 μg bulk BeO powder) onto Whatman 41 (42.5 mm diameter) cellulose filters (catalog number 1441-042,

Whatman International Ltd., Maidstone, England).

Each spiked filter sample (n = 1 at 6 different mass levels) was hot-plate digested with 1 mL 50 % sulfuric acid and 1 mL concentrated nitric acid with addition of 30 % hydrogen peroxide until a white vapor was produced. The solution was cooled, 1 mL hydrochloric acid added, and then reheated to near boiling. Next, the solution was diluted to 10 mL with ASTM Type I water (resultant matrix 10 % hydrochloric acid, 4 % sulfuric acid) and analyzed by inductively coupled plasma atomic emission spectroscopy (ICP-AES) (Optima 4300DV, Perkin Elmer) at the 313.107 nm beryllium emission line.

Modified NIOSH Method 7300—Recovery of beryllium, from BeO powder digested by a modified NIOSH Method 7300, was performed by a private laboratory (different from the laboratory that performed analyses according to OSHA Method 125G). Spiked samples were prepared for the purpose of digesting and analyzing samples known to contain large quantities of organic material. Method 7300 was modified by using large quantities of acid (40 to 60 mL nitric, 5 mL perchloric acid) to fully digest thin cotton gloves (Lisle 3301, Johnson Wilshire Inc., Downey, CA) spiked with BeO. A suspension of known concentration was prepared by adding 10.2 mg bulk BeO powder to 1 L n-propanol, then subjecting to ultrasonic sonication for 30 min. To prepare a spike sample, a known volume of suspension that contained 10.2 μg BeO was pipetted onto a cotton glove. One mL aliquots of the suspension itself were also analyzed.

Each spiked cotton glove sample (n = 6) and each suspension sample (n = 5) was hot-plate digested in a beaker with 25 mL concentrated nitric acid at 100°C for 2 h to break the integrity of the glove matrix. Two mL of perchloric acid were then added to each beaker, the beaker covered, and the sample refluxed at 150°C for 48 h. The cover was then removed and the sample taken to dryness at the same temperature. This digestion procedure was repeated with two additional aliquots of nitric acid (5 mL) and perchloric acid (1 mL). The resultant residue was dissolved in 25 mL of 4 % nitric/1 % perchloric acid, filtered through a 0.45 μm pore size polytetrafluoroethylene filter, and analyzed by ICP-AES (Spectro EOP, Spectro Analytical Instruments, Kleve, Germany) at the 313.107 nm beryllium emission line. All sample results were background-corrected for levels of beryllium in propanol and in the cotton glove matrix.

EPA Method 3015—Recovery of beryllium from the soluble beryllium standard reference material digested by EPA Method 3015 was performed by the same commercial laboratory as described for modified EPA Method 3051. Liquid spike samples (n = 77) were prepared gravimetrically (0.1 to 25 μg beryllium) from a working solution of liquid beryllium standard reference material (SRM3105a, National Institute of Standards and Technology, Gaithersburg, MD) in phagolysosomal simulant fluid [8] and submitted blind to the laboratory for quantification. For each aqueous sample, 45 mL of well-mixed sample were measured into a clean microwave vessel. Five milliliters of high-purity concentrated nitric acid were added and the vessel swirled to mix. Vessels were sealed and heated in a microwave for 30 min, cooled, transferred to 50 mL centrifuge cones, and analyzed without further dilution by ICP-MS.

Results

BeO and Beryllium Metal Properties

Morphology and size of UOX-125 BeO powder and I-400 metal powder collected in stage 1 and 5 of the aerosol cyclone are shown in Fig. 1 (note that the scale used for the BeO powder images differed from the scale used for the beryllium metal powder images, precluding direct comparison of particle size between materials). UOX-125 BeO powder had aggregate cluster morphology; the average primary particle size, 0.19 ± 0.42 μm, was independent of cluster size. Metal powder had compact morphology and size that decreased with aerodynamic cutoff diameters of the cyclone. The SSA of the BeO was independent of aerodynamic particle cluster size, but dependent upon average BeO primary particle size, while SSA of metal powder increased as aerodynamic particle size decreased (Fig. 2).

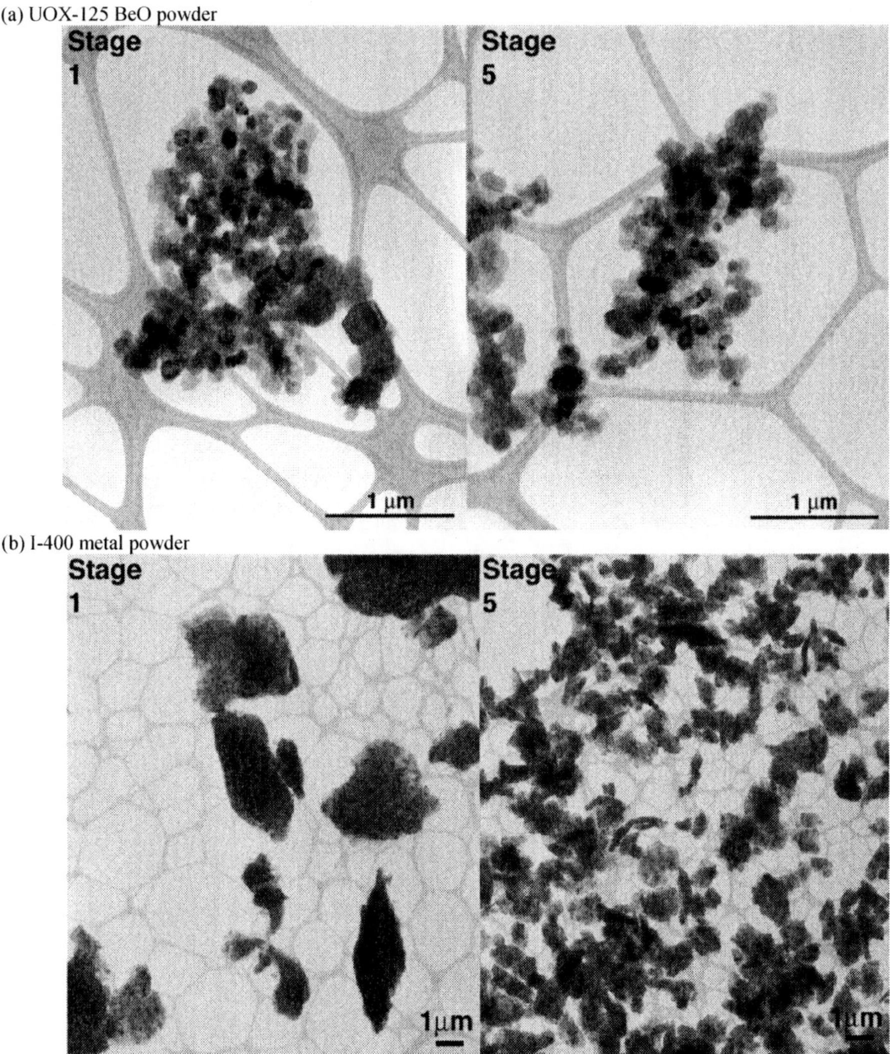

FIG. 1—*Transmission electron micrographs of* (a) *UOX-125 BeO powder and* (b) *I-400 metal powder collected in stage 1 and stage 5 of the aerosol cyclone. The micrographs illustrate that BeO consists of agglomerate clusters of uniform diameter primary particles (0.19 ± 0.42 μm), and metal powder consists of compact particles of decreasing size.*

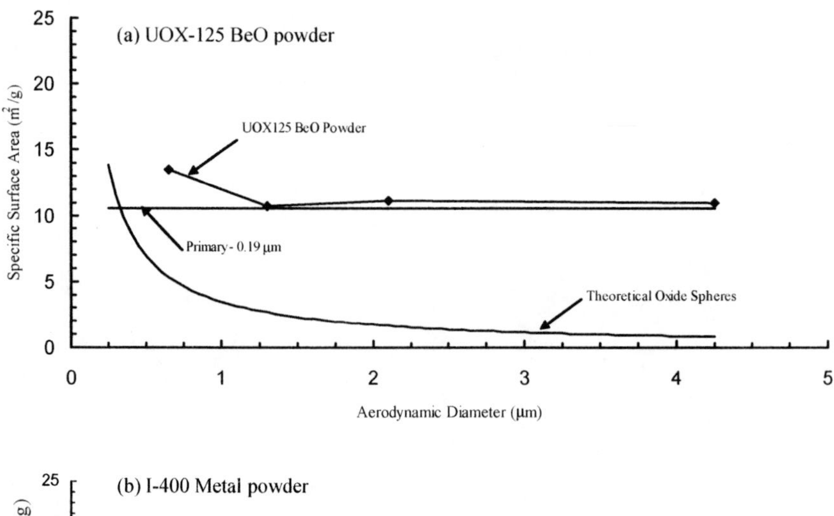

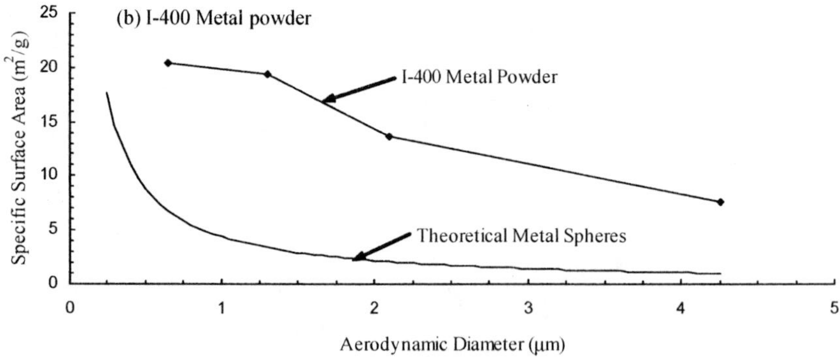

FIG. 2—*Specific surface area (SSA) of* (a) *UOX-125 BeO powder is independent of cluster size but depends on primary particle size. SSA of* (b) *I-400 metal powder is dependent on particle size.*

Analysis of crystalline composition by XRD, TEM-SAD, and TEM-µD identified only BeO in UOX-125 BeO powder and only beryllium metal in I-400 metal powder (i.e., both materials were high-purity). Analysis of elemental composition by TEM-EDS and TEM-EELS identified beryllium, oxygen, and silicon in both powders. X-ray photoelectron spectroscopy analysis identified only BeO on the surface of the BeO powder. Beryllium oxide and beryllium metal were identified on the surface of the metal powder. From the XPS data, the thickness of the BeO layer on the surface of metal powder was estimated to be 42 Å thick.

Powder Dissolution Behavior

Values of k ($g \cdot cm^{-2} \cdot day^{-1}$) determined for BeO powder and metal powder in 0.1 N hydrochloric acid, phagolysosomal simulant fluid, and serum ultrafiltrate are summarized in Table 2. The k values for BeO powder were about a factor of 10 lower than for metal powder in acidic solvents. Figure 3 is a plot of both the theoretical initial fractional particle dissolution rate and particle dissolution lifetime for compact smooth spheres of BeO powder (density 3.0 g•cm^{-3}) having initial diameters (D_0) of 0.1, 1, 10, and 100 μm and k values ranging from 10^{-10} to 10^{-1} g•cm^{-2}•day^{-1}. In accordance with Eq 1 and Eq 2, dissolution rate and dissolution lifetime are observed to vary proportionally with particle diameter.

TABLE 2—*Estimated values of k for BeO powder and metal powder in 0.1 N hydrochloric acid, phagolysosomal simulant fluid (pH 4.5), and simulated lung fluid (pH 7.3).*

Powder	Solvent[A]	pH	k, g·cm^{-2}·day^{-1}
UOX-125 BeO	HCl	1	$6.1 \pm 2.2 \times 10^{-8}$
	PSF	4.5	$1.2 \pm 1.4 \times 10^{-8}$
	SUF	7.3	$3.7 \pm 1.2 \times 10^{-9}$
I-400 metal	HCl	1	$4.1 \pm 0.2 \times 10^{-7}$
	PSF	4.5	$1.1 \pm 1.4 \times 10^{-7}$
	SUF	7.3	$1.5 \pm 0.8 \times 10^{-9}$

[A] HCl = hydrochloric acid [7].
PSF = phagolysosomal simulant fluid (model of pulmonary alveolar macrophage phagolysosome fluid) [9].
SUF = serum ultrafiltrate (model of extracellular lung fluid) [7].

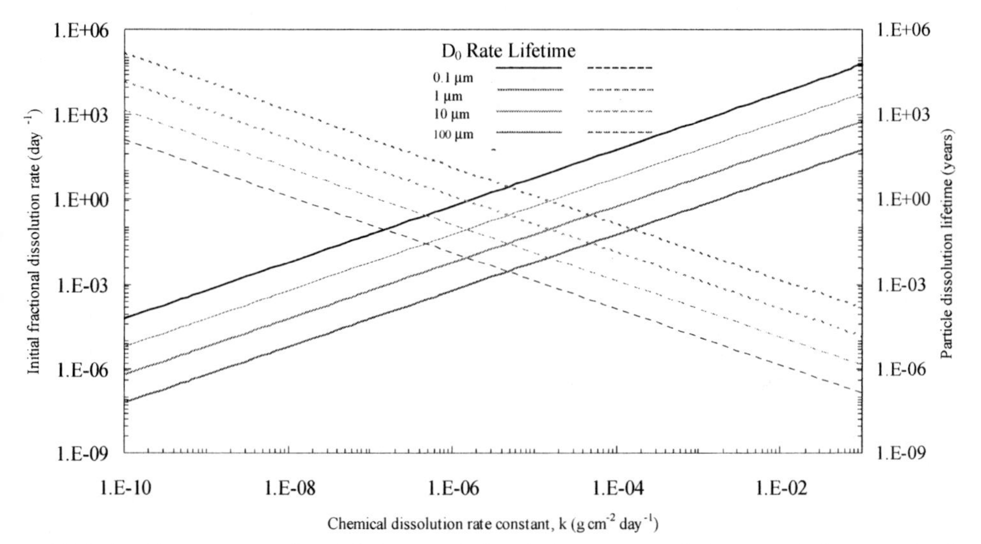

FIG. 3—*Comparison of initial fractional dissolution rates and times needed to completely dissolve BeO particles having initial particle diameters of 0.1, 1, 10, and 100 μm and chemical dissolution rate constants (k) of 10^{-10} to 10^{-1} g•cm^{-2}•day^{-1}.*

Digestion and Recovery Efficiencies

Digestion and recovery efficiencies for the beryllium powders by commonly used standard analytical methods are summarized in Table 3. Greater than 90 % of BeO was digested and recovered by EPA Method 3051 and NIOSH Method 7300. In contrast, less than 77 % of BeO was digested and recovered by OSHA Method 125G. For I-400 metal powder, 89 % (coefficient of variation, CV = 8.9 %) was digested and recovered by EPA Method 3051. Recovery of the aqueous standard reference material digested by modified EPA Method 3015 was complete (100.4 %) but tended to be more variable (CV = 14 %) than observed for powder samples.

TABLE 3—*Recovery of beryllium from UOX-125 BeO and I-400 metal powders digested and analyzed using modified versions of standard analytical methods.*

Modified Method[A]	Treatment	Powder	N	Recovery Mean (CV)[B], %
EPA 3051	Microwave	Metal	34	88.6 (8.9)[C]
EPA 3051	Microwave	BeO	49	93.7 (9.8)[C]
OSHA 125G	Hot plate	BeO	6	76.6 (5.7)[D]
NIOSH 7300	Hot plate	BeO	5	95.9 (1.0)[E]
NIOSH 7300	Hot plate	BeO	6	94.6 (0.9)[F]

[A] Standard analytical methods were modified as follows:

EPA 3051: A 4:1 concentrated HNO_3 to concentrated HCl solution, rather than 10 mL concentrated HNO_3 was used to digest the sample; sample solution was microwave-heated for 30 min, rather than 10 min.

OSHA 125G: A 2.5:1 50 % H_2SO_4 to concentrated HNO_3 solution, rather than a 1:2 ratio of these acids was used to ash the sample; the sample solution was placed on a hotplate straight away rather than let sit for 1 h at ambient temperature; and 1 mL rather than 4 mL concentrated HCl was used to reheat the sample to near boiling.

NIOSH 7300: 25 mL HNO_3, rather than 4:1 HNO_3 to $HClO_4$ solution was used to ash; the sample solution was placed on a hotplate straight away, rather than let sit at ambient temperature for 0.5 h; 2 mL $HClO_4$, rather than 4:1 HNO_3 to $HClO_4$ solution was added to the ashing solution; this solution was refluxed for 48 h, then taken to dryness; next 5:1 HNO_3 to $HClO_4$ was added to the solution and refluxed to near dryness twice more, rather than adding 4:1 HNO_3 to $HClO_4$ solution and heating repeatedly until the solution turns clear; finally, solid debris from cotton gloves were removed by filtering digestate through a 0.45 μm pore size polytetrafluoroethylene filter.

[B] CV = coefficient of variation.

[C] Powder suspended in PBS and spiked onto 37-mm diameter cellulose filter support pads.

[D] BeO powder suspended in PBS and spiked onto 42.5-mm diameter Whatman cellulose filters.

[E] BeO powder suspended in n-propanol.

[F] BeO powder suspended in n-propanol and spiked onto cotton gloves.

Discussion

Fundamental to the assessment of the performance of any analytical method is the existence of accurate standard reference materials. Standard reference materials must be: 1) independently validated, 2) representative of both the chemical form of the material in the actual samples and matrices in which that material will be found, and 3) sufficiently robust to ensure complete digestion of the most insoluble chemical form. We used a suite of analytical techniques to assess the properties of particulate BeO and beryllium metal powder materials and analyzed the same materials by spectroscopic analysis. Two forms of beryllium with toxicological significance were evaluated for a range of sample matrices, including suspension of particles in n-propanol or PBS and application to mixed cellulose ester filters or thin cotton gloves.

The UOX-125 BeO powder had aggregate cluster morphology with SSA independent of particle cluster size. In turn, dissolution kinetics of the BeO powder was independent of cluster size but dependent on primary particle size (0.19 μm) [9]. However, to fully evaluate the robustness of a digestion procedure for BeO, compact particles from machining or other comminution of sintered BeO that have a range of physical particle sizes are also needed. I-400 beryllium metal powder had compact particle morphology with SSA that increased as particle size decreased, which, as previously shown [7], caused size-dependent dissolution kinetics (i.e., smaller particles dissolved more quickly than larger particles). Note that digestion of bulk samples of metal powder may differ from that observed for the size-selective materials (<10 μm aerodynamic diameter) used in our studies. I-400 metal powder consisted of particles that had passed through a 400 mesh screen (i.e., nominal particle diameter less than 38 μm). Beryllium metal particles collected on substrate during air or swipe sampling could contain particles with diameters larger than 38 μm. In this case, a standard reference material of metal powder with particle diameter greater than 38 μm would be needed to fully evaluate the robustness of a digestion procedure. Successful digestion of metal particles of these sizes would ensure complete digestion of smaller metal particles.

Both beryllium powder materials were high purity. The estimated oxide layer thickness on the surface of metal powder (42 Å) was in excellent agreement with 52 Å that was previously determined for this same size of metal powder using NAA [13]. Note that there is a significant difference in dissolution of BeO and metal even though the metal powder had a 42 Å oxide coating (Table 2). This difference in k values has implications for complete digestion of beryllium-containing particles, which may include intentional oxides (e.g., UOX-125 BeO powder) or incidental oxides (e.g., BeO formed under ambient conditions on the surface of a beryllium material). Note that for intentional oxides, calcine (heat-treatment) temperature influences particle physicochemical properties, which in turn influences solubility [7]. Given the k values in acid solvents (Table 2), the time required to completely dissolve a particle of intentionally formed UOX-125 BeO is more than a factor of 10 longer than for the same size particle of I-400 beryllium metal that has a thin surface coating of incidental oxide (Eq 2). A given k value for digestion is related to the chemical properties of the particle, the chemical activity of the solvent, and thermal conditions of the digestion procedure. Thus, it is not just the chemical form of material and the solvent or solvent/temperature combination used in the digestion procedure that determines digestion efficiency; time is also a factor. For a given chemical activity and set of thermal conditions of a digestion procedure, the larger the particle, the longer the time needed for complete digestion. Digestion procedures used in EPA Method 3051, NIOSH Method 7300, and OSHA Method 125G may dissolve nanometer-scale beryllium-containing particles, but above a certain particle size (which is currently not known), digestion is not complete. The time required to completely dissolve the largest particle of a given chemical form within a sample (and all particles of the same chemical form of smaller sizes), under the solvent and temperature conditions of a given digestion procedure, can be estimated using Eqs 1 and 2. UOX-125 BeO particles with diameters equal to or less than 100 μm are completely dissolved in 1.2 h when $k = 10^{-1}$ $g \cdot cm^{-2} \cdot day^{-1}$. When $k = 10^{-5}$ $g \cdot cm^{-2} \cdot day^{-1}$, UOX-125 BeO particles with diameter equal to or less than 100 μm are completely dissolved in 12 000 h (Fig. 3). In general, the time needed to completely dissolve a particle of a given size and chemical form will increase by an order of magnitude for each order of magnitude lowering of k value.

The ranking of digestion efficiency for beryllium from UOX-125 BeO by method was (from highest to lowest recovery): NIOSH Method 7300 ≈ EPA Method 3051 > OSHA Method 125G. Many factors can influence the digestion efficiency of these analytical methods, including the suspension preparation technique and the selection of sample media, mass level, and digestion procedure. Variability in the chemical composition and particle size of the BeO digested can be excluded as factors to explain differences in digestion efficiency among these common analytical methods.

Suspensions analyzed by EPA Method 3051 and OSHA Method 125G were prepared by the same person in the same laboratory under similar conditions of temperature and low humidity (note that BeO and metal are not hygroscopic), and techniques for agitation of suspensions and dispensing onto sample media were, by design, very similar. PBS was used to suspend BeO for analysis by EPA Method 3051 and OSHA Method 125G, while n-propanol was used to suspend particles for analysis by NIOSH 7300. Phosphates in PBS could cause spectral interference, but similar levels of beryllium were recovered by EPA Method 3051 and NIOSH Method 7300. These data indicate that variation in suspension preparation techniques may explain a portion, but not all, of the observed difference in beryllium recovery among techniques.

Cellulose filters were used to prepare spike samples for analysis by EPA Method 3051, Whatman cellulose filters were used to prepare spike samples for analysis by OSHA Method 125G, while cotton gloves were used as the sample media to prepare spike samples for analysis by NIOSH Method 7300. Beryllium recovery from EPA Method 3051 was similar to NIOSH 7300, indicating that differences between the spike sample media probably do not fully explain the low recovery by OSHA Method 125G.

Beryllium oxide mass levels were highest on spike samples digested by EPA Method 3051 (50 to 850 μg BeO) and were similar for spike samples analyzed by NIOSH Method 7300 (10.2 μg BeO) and OSHA Method 125G (0.05 to 10.0 μg BeO). Despite the fact that BeO masses differed by up to a factor of 80 between spiked samples digested by EPA Method 3051 and NIOSH Method 7300, beryllium recovery levels were proportionally similar. The low recovery of beryllium from BeO by OSHA Method 125G cannot be explained by differences between spike sample mass levels. Note that the influence of mass level on spike sample recovery by OSHA Method 125G could be assessed by successive digestions (e.g., digestion for 10 min, 100 min, and 1000 min). Particles would be completely dissolved by the method when element level results remained unchanged between successive digestion durations.

The relatively small contributions of suspension preparation techniques, sample media, and sample mass levels to observed differences in recovery efficiencies suggest that differences in digestion procedures used by EPA Method 3051, NIOSH Method 7300, and OSHA Method 125G were important for dissolution of beryllium particles. Our data indicate that the use of perchloric acid or a microwave digestion step was needed to nearly completely dissolve 0.2-μm diameter BeO primary particles. Note that microwave digestion did not completely dissolve all sizes of metal powder in samples. With respect to UOX-125 BeO, even though a digestion procedure can completely dissolve 0.2 μm diameter BeO particles, it may not completely dissolve a larger (e.g., 1 μm) BeO particle. For larger BeO particles, a longer digestion time may be needed to completely dissolve the particles prior to measurement. Note that other digestion procedures, e.g., ASTM D 7035-04 [22] or International Standards Organization 15202-2 [23], and other digestion solvents (e.g., hydrofluoric acid) exist but were not evaluated in this study to determine if they could completely dissolve BeO and metal particles.

Beryllium samples for proficiency testing programs are often prepared using a soluble beryllium acetate standard reference material, rather than a poorly soluble particulate beryllium metal and BeO powder. Analytical laboratory methods currently considered proficient in beryllium analysis based on the results of proficiency testing using beryllium acetate may, in fact, not be completely digesting particulate BeO and metal. A test program to evaluate the robustness of the digestion method and the accuracy of the instrument analysis would include a suite of standard reference materials consisting of soluble beryllium salts and poorly soluble particulate beryllium metal and oxide in a range of particle sizes.

Implications of Lower-Than-Expected Recovery Efficiency

If beryllium recovery efficiency by an analytical method is not accurately known, routine analysis of environmental samples may be in error when estimating beryllium levels in samples.

For example, if complete recovery by a method was assumed, but actual recovery was 77 %, then the estimated beryllium levels may be lower than actual levels. Our results indicate that spike samples can be prepared from suspensions of beryllium powder in PBS or n-propanol and used for blind assessment of method digestion and recovery efficiency. Note that use of the UOX-125 BeO powder or I-400 metal powder as a reference material cannot help to speciate (identify) the chemical form(s) of beryllium that may be present in an environmental sample. Speciation can be important for discriminating forms of beryllium in the workplace and can aid in identifying sources of exposure. Speciation can also help discriminate between beryllium from naturally occurring minerals in the soil and beryllium from man-made sources. Various techniques to speciate forms and sources of beryllium have been proposed, including serial digestion [24] and elemental ratios in local soils [25,26].

Summary and Recommendations

We characterized two candidate analytical reference materials for analytical chemistry and occupational health studies: aerodynamically size-separated type UOX-125 BeO powder and type I-400 beryllium metal powder. Each high-purity material presented a unique challenge for digestion. Beryllium oxide powder has aggregate cluster morphology with SSA independent of particle cluster size, resulting in dissolution kinetics that is independent of aerodynamic cluster size. I-400 beryllium metal powder has compact particle morphology with SSA that increased as particle size decreased, resulting in size-dependent dissolution kinetics. Use of the BeO material in evaluation of the recovery efficiencies of 3 United States governmental standard analytical methods demonstrated that not all methods provide quantitative recovery for BeO particles. Perchloric acid or a microwave-assisted digestion nearly completely dissolved 0.2 μm diameter BeO primary particles. Microwave-assisted digestion did not completely dissolve all sizes of metal powder in spiked samples.

Based on the results of this study, the following recommendations are made:

- In addition to the chemical form of beryllium, a sample preparation procedure must consider the digestion solvent, solvent temperature, particle size, and time.
- Mercer single-particle dissolution theory could be used to determine the time needed to completely dissolve the largest particle in a sample (Eq 2).
- To fully evaluate the robustness of a digestion procedure, reference materials of compact particles from machining or other comminution of sintered BeO that have a range of physical particle sizes are needed, as are particles of metal powder with particle diameter greater than 38 μm.
- Upon being validated, BeO powder and metal powder standard reference materials should be incorporated into proficiency testing programs.
- Laboratory collaborations should continue to evaluate (e.g., round-robin testing) and apply these BeO and beryllium metal powders as analytical reference materials.

Acknowledgments

The authors thank R. Dickerson, E. Peterson, and R. Schulze (LANL) for collaboration on the microanalyses; M. Millson (NIOSH-DART) for performing ICP-AES analyses; and J. Burkhart, C. Coffey, K. Kreiss, T. Pearce, and C. Schuler (NIOSH) for their useful review and discussion of this manuscript. This project was supported by NIOSH Research Grant 1R03

OH007447-01 and by the LANL HSR Division TDEA program. A. B. Stefaniak received support from the NIEHS Training Program in Environmental Health Sciences ES07141.

References

[1] Sterner, J. H. and Eisenbud, M., "Epidemiology of beryllium intoxication," *A.M.A. Archives of Industrial Hygiene and Occupational Medicine*, Vol. 4, No. 1, 1951, pp. 123–151.
[2] Kreiss, K., Mroz, M., Newman, L. S., Martyny, J., and Zhen, B., "Machining risk of beryllium disease and sensitization with median exposures below 2 $\mu g/m^3$," *American Journal of Industrial Medicine*, Vol. 30, 1996, pp. 16–25.
[3] Kreiss, K., Mroz, M., Zhen, B., Wiedemann, H., and Barna, B., "Risks of beryllium disease related to work processes at a metal, alloy, and oxide production plant," *Occupational and Environmental Medicine*, Vol. 54, No. 8, 1997, pp. 605–612.
[4] Eisenbud, M., "The standard for control of chronic beryllium disease," *Applied Occupational and Environmental Hygiene*, Vol. 13, No. 1, 1998, pp. 25–31.
[5] Maier, L. A. and Newman, L. S., "Beryllium Disease," *Environmental and Occupational Medicine*, Rom, W. N., Ed., Philadelphia, Lippincott-Raven, 1998, pp. 1017–1031.
[6] Henneberger, P. K., Goe, S. K., Miller, W. E., Doney, B., and Groce, D. W., "Industries in the United States with airborne beryllium exposure and estimates of the number of current workers potentially exposed," *Journal of Occupational and Environmental Hygiene*, Vol. 1, No. 10, 2004, pp. 648–659.
[7] Finch, G. L., Mewhinney, J. A., Eidson, A. F., Hoover, M. D., and Rothenberg, S. J., "*In vitro* dissolution characteristics of beryllium oxide and beryllium metal aerosols," *Journal of Aerosol Science*, Vol. 19, No. 3, 1988, pp. 333–342.
[8] Stefaniak, A.B., Guilmette, R.A., Day, G.A., Hoover, M.D., Breysse, P.N., and Scripsick, R.C.: Characterization of simulated phagoloysosomal fluid for study of beryllium dissolution. *Toxicology In Vitro*. 19(1):123–134 (2005).
[9] Stefaniak, A. B., Day, G. A., Hoover, M. D., Breysse, P. N., and Scripsick, R. C., "Differences in dissolution behavior in a phagolysosomal simulant fluid for single-constituent and multi-constituent materials associated with beryllium sensitization and chronic beryllium disease," (submitted).
[10] Stange, A. W., Hilmas, D. E., and Furman, F. J., "Possible health risk from low-level exposure to beryllium," *Toxicology*, Vol. 111, 1996, pp. 213–224.
[11] Henneberger, P. K., Cumro, D., Deubner, D. D., Kent, M. S., McCawley, M., and Kreiss, K., "Beryllium sensitization and disease among long-term and short-term workers in a beryllium ceramics plant," *International Archives of Occupational and Environmental Health*, Vol. 74, 2001, pp. 167–176.
[12] Kelleher, P. C., Martyny, J. W., Mroz, M. M., Maier, L. A., Ruttenber, A. J., Young, D. A., et al., "Beryllium particulate exposure and disease relations in a beryllium machining plant," *Journal of Occupational and Environmental Medicine*, Vol. 43, No. 3, 2001, pp. 238–249.
[13] Hoover, M. D., Castorina, B. T., Finch, G. L., and Rothenberg, S. J., "Determination of the oxide layer thickness on beryllium metal particles," *American Industrial Hygiene Association Journal*, Vol. 50, No. 10, 1989, pp. 550–553.
[14] Stefaniak, A B., Hoover, M. D., Dickerson, R. M., Peterson, E. J., Day, G. A., Breysse, P. N., et al., "Surface area of respirable beryllium metal, oxide, and copper alloy aerosols and implications for assessment of exposure risk of chronic beryllium disease," *American Industrial Hygiene Association Journal*, Vol. 64, 2003, pp. 297–305.
[15] Stefaniak, A. B., Hoover, M. D., Day, G. A., Dickerson, R. M., Peterson, E. J., Kent, M. S., et al, "Characterization of physicochemical properties of beryllium aerosols associated with chronic beryllium disease," *Journal of Environmental Monitoring*, Vol. 6, No. 6, 2004, pp. 523–532.
[16] Smith, W. B., Wilson, R. R., and Harris, D. B., "A five-stage cyclone system for *in situ*

sampling," *Environmental Science and Technology*, Vol. 13, No. 11, 1979, pp. 1387–1392.
[17] Kanapilly, G. M., Raabe, O. G., Go, C. H. T., and Chimenti, R. A., "Measurement of *in vitro* dissolution of aerosol particles for comparison to *in vivo* dissolution in the lower respiratory tract after inhalation," *Health Physics*, Vol. 24, 1973, pp. 497–507.
[18] Mercer, T. T., "On the role of particle size in the dissolution of lung burdens," *Health Physics*, Vol. 13, 1967, pp. 1211–1221.
[19] United States Environmental Protection Agency. SW-846: Test Methods for Evaluating Solid Waste, Physical/Chemical Methods. US EPA Office of Solid Waste. Update III. NTIS. Springfield, VA, 1996.
[20] United States Occupational Health & Safety Administration. Method Number ID-125G, "Metal and Metalliod Particulates in Workplace Atmospheres (ICP Analysis)," http://www.osha.gov/dts/sltc/methods/inorganic/id125g/id125g.html, (accessed March, 2004).
[21] National Institute for Occupational Safety and Health, 1994, NIOSH Manual of Analytical Methods (NMAM®), 4th ed., DHHS (NIOSH) Publication 94-113 (August, 1994), 1st Supplement Publication 96-135, 2nd Supplement Publication 98-119, 3rd Supplement 2003-154, Schlecht, P.C. & O'Connor, P.F., Eds. www.cdc.gov/niosh/nmam.
[22] ASTM Standard D 7035, "Standard Test Method for Determination of Metals and Metalloids in Airborne Particulate Matter by Inductively Coupled Plasma Atomic Emission Spectrometry (ICP-AES)," ASTM International, West Conshohocken, PA, 2004.
[23] International Standards Organization Method 15202-2, "Workplace air - Determination of metals and metalloids in airborne particulate matter by inductively coupled plasma atomic emission spectrometry -- Part 2: Sample preparation," ISO, Geneva, Switzerland, 2001.
[24] Profumo, A., Spini, G., Cucca, L., and Pesavento, M., "Determination of inorganic beryllium species in the particulate matter of emissions and working areas," *Talanta*, Vol. 57, 2002, pp. 929–934.
[25] Myers, J. and Thorbjornsen, K., "Identifying metals contamination in soil: A geochemical approach," *Soil & Sediment Contamination*, Vol. 13, 2004, pp. 1–16.
[26] Longmire, P. A., Reneau, S. L., Watt, P. M., McFadden, L. D., Gardner, J. N., Duffy, C. J., et al., "Natural Backround Geochemistry, Geomorphology, and Pedogenesis of Selected Soil Profiles and Bandelier Tuff, Los Alamos, New Mexico," Los Alamos National Laboratory Report LA-12913-MS, UC-903, May 1996.

Journal of ASTM International, November/December 2005, Vol. 2, No. 10
Paper ID JAI13166
Available online at www.astm.org

Shakker Amer, Ph.D.,[1] *Doug Smieja,*[2] *Jason Loughrin,*[3] *and Lyle Reichmann, CIH*[4]

Determination of Beryllium Compounds by NIOSH 7303

ABSTRACT: The fourth edition of the National Institute for Occupational Safety and Health (NIOSH) Manual of Analytical Methods contains a new, simplified procedure for elemental analysis in workplace atmospheres by Inductively Coupled Argon Plasma Atomic Emission Spectroscopy (ICP-AES) titled NIOSH 7303. This method uses hot block digestion with hydrochloric acid (HCl) and nitric acid (HNO_3). The method was presented as fully valid for elemental beryllium. It was further investigated and also found to be valid for beryllium sulfate tetrahydrate and has a recovery efficiency of about 80 % for beryllium oxide. Additionally, no matrix effect on beryllium analysis is apparent by NIOSH 7303 for five common sampling media: polyvinylchloride filters, mixed cellulose ester filters, Whatman® 42 filters, Ghost™ wipes, and Palintest® wipes. Finally, the method has been reevaluated and found to produce a beryllium detection level of 0.0002 µg/mL (0.005 µg/sample for air samples) and a quantifiable level of 0.0005 (0.0125 µg/sample). At this detection limit, the Occupational Safety and Health Administration Permissible Exposure Limit (PEL) of 0.002 mg/m^3 can be achieved at the method's detection level with an air sample of 2.5 L or one tenth of the PEL with a sample of 25 L under the prescribed preparation and analytical conditions.

KEYWORDS: beryllium, beryllium oxide, ICP, NIOSH 7303

Introduction

Beryllium (Be) is found in a variety of different forms including metal, soluble salts, alloys, and oxide. Its properties have made it indispensable in modern high technology materials including aerospace materials, nuclear materials, and metal alloys [1]. Unfortunately, it is also associated with a number of toxic effects, such as acute and chronic beryllium disease [1,2].

Different methods have been developed to extract beryllium metal, beryllium compounds, and beryllium oxide using perchloric acid ($HClO_4$), hydrogen fluoride (HF), hydrogen peroxide (H_2O_2), and sulfuric acid (H_2SO_4) heated until it fumes [3–5]. These acids may require special safety precautions during digestion, or they may create viscosity and inter-element interference issues with ICP-AES analysis.

The National Institute for Occupational Safety and Health's (NIOSH) Manual of Analytical Methods (NMAM) 7303 uses only hydrochloric acid (HCl) and nitric acid (HNO_3) and was previously validated for elemental beryllium [6]. This method was recently evaluated for beryllium sulfate tetrahydrate ($BeSO_4 \cdot 4H_2O$) and beryllium oxide (BeO). Matrix effects caused by different sampling media have been examined for polyvinylchloride (PVC) filters, mixed cellulose ester (MCE) filters, Whatman® 42 filter paper, Ghost™ wipes, and Palintest® wipes. The detection level was also reevaluated and found to be 0.0002 micrograms per milliliter (µg/mL) or 0.005 micrograms per sample (µg/sample) and a quantifiable limit of 0.0005 µg/mL

Manuscript received 23 February 2005; accepted for publication 6 May 2005; published November 2005. Presented at ASTM Symposium on Beryllium: Sampling and Analysis on 21-22 April 2005 in Reno, NV.
[1] Senior Chemist, Wisconsin Occupational Health Laboratory, 2601 Agriculture Dr., Madison, WI 53718.
[2] Advanced Chemist, Wisconsin Occupational Health Laboratory, 2601 Agriculture Dr., Madison, WI 53718.
[3] Senior Chemist, Wisconsin Occupational Health Laboratory, 2601 Agriculture Dr., Madison, WI 53718.
[4] Chemist Supervisor, Wisconsin Occupational Health Laboratory, 2601 Agriculture Dr., Madison, WI 53718.

or 0.0125 μg/sample using a 25 mL dilution volume. At this detection level, the Occupational Safety and Health Administration's (OSHA) Permissible Exposure Limit (PEL) of 0.002 milligrams per cubic meter (mg/m^3) can be achieved with an air volume of 2.5 L, and one tenth of the PEL can be achieved with a 25-L air sample. An air volume of 25 L will meet the proposed American Conference of Governmental Industrial Hygienist (ACGIH) Threshold Limit Value (TLV) of 0.0002 mg/m^3, and an air volume of 250 L will meet one tenth of the TLV. An occupation exposure limit of 0.0001 mg/m^3 was recently advocated by individuals from the U.S. Department of Energy (DOE) [7]. This value can be achieved with an air sample of 50 L, while a 500-L air volume is required to achieve one tenth of the exposure limit.

Experimental

Nine samples of BeO (Aldrich, 99.98 %, catalog number 202770-5G) were prepared with weights ranging from 3.88–52.55 milligrams (mg) of BeO using an analytical balance. The samples were digested and analyzed using NIOSH NMAM 7303 with an inductively coupled argon plasma atomic emission spectrometer (ICP-AES). All samples were analyzed on a Thermo Jarrell Ash 61E radial simultaneous plasma emission spectrometer at a wavelength of 313.042 nanometers (nm). Samples were run three times starting with the day of digestion (day 0, day 1, and day 7). Results were compared to the known weighed amounts to determine the validity of the method for extracting and analyzing BeO in bulk samples and air samples. Nine samples of $BeSO_4 \cdot 4H_2O$ (Aldrich, 99.99 %, catalog number 202789-50G) were also prepared with weights ranging from 4.67–51.37 mg and analyzed in the same manner.

Matrix effects from sampling media were investigated using 37 mm 0.8 micron (μm) MCE filters (Zefon catalog number FMCE837, lot number 5786), 37 mm 0.8 μm PVC filters (catalog number FPVC537, lot number 6006), 90 mm Whatman® 42 filter paper (lot number 928913), Ghost™ wipes (manufacture date 08/01/02), and Palintest® wipes (manufacture date 08/02). Various amounts of beryllium liquid standard (SPEX 1000 μg/mL, catalog number PLBE2-2X) at 0.001, 0.01, 0.1, 1.0, 10, and 20 μg/mL final sample concentration were spiked in triplicate on each matrix. MCE and PVC filters were digested as described in NIOSH 7303 and had total volumes of 25 mL. Wipe materials were digested by doubling the amount of acids at each step and brought to a final volume of 50 mL.

The method detection and quantitation levels (MDL, MQL) were investigated by initially running seven blank samples for each matrix to estimate the MDL and MQL using the student t-value of 3.14 times the standard deviation of the replicates. Then seven liquid spikes using beryllium standard at approximately five times the highest estimated MDL (Ghost™ wipes) were analyzed for each matrix to determine the actual MQL and MDL using the same lot numbers listed above. Again, MCE and PVC filters had total volumes of 25 mL, and wipe materials had total volumes of 50 mL.

Although it was not statistically significant, the original validation data showed a possible reduction in recovery over the seven-day investigation. Sample stability was reinvestigated using three samples each of Be (Aldrich, 325 mesh, 99+ %, catalog number 378135-5G), BeO, and $BeSO_4 \cdot 4H_2O$ and run on day 0, day 1, and day 7.

Results

A summary of percent recovery of BeO and $BeSO_4 \cdot 4H_2O$ is listed in Tables 1 and 2, respectively. The average recovery of the BeO samples on day 0 was 83.0 % with a range of

71.9–87.6 % and a standard deviation of 5.78 (two of the samples were not run on day 0). On day 1 the average was 81.1 % with a range of 78.8–85.0 % and a standard deviation of 1.83. On day seven the average recovery was 78.0 % with a range of 74.0–79.7 % and a standard deviation of 1.88.

TABLE 1—*BeO validation.*

Sample	Weight (g)	Result (g) day 0	Recovery day 0	Result (g) day 1	Recovery day 1	Result (g) day 7	Recovery day 7
1	0.00388	0.0034	86.4 %	0.0033	85.0 %	0.0031	78.9 %
2	0.00975	0.0085	87.6 %	0.0079	81.0 %	0.0078	79.7 %
3	0.01515	0.0131	86.1 %	0.0123	81.2 %	0.0120	79.5 %
4	0.02153	0.0188	87.2 %	0.0178	82.6 %	0.0170	78.9 %
5	0.02852	0.0237	82.9 %	0.0230	80.7 %	0.0225	79.0 %
6	0.03383	0.0267	78.8 %	0.0269	79.5 %	0.0266	78.5 %
7	0.03950	0.0284	71.9 %	0.0319	80.7 %	0.0302	76.4 %
8	0.04599	NA (a)	NA (a)	0.0369	80.2 %	0.0353	76.7 %
9	0.05255	NA (a)	NA (a)	0.0414	78.8 %	0.0389	74.0 %
Average Recovery			83.0 %		81.1 %		78.0 %
Standard Deviation			5.78		1.83		1.88

(a) N.A. – not applicable because samples were not run on day 0.

TABLE 2—*$BeSO_4 \cdot 4H_2O$ validation.*

Sample	Weight (g)	Result (g) day 0	Recovery day 0	Result (g) day 1	Recovery day 1	Result (g) day 7	Recovery day 7
1	0.00467	0.0049	103.8 %	0.0051	109.2 %	0.0047	100.5 %
2	0.01035	0.0107	103.1 %	0.0111	106.9 %	0.0103	99.7 %
3	0.01674	0.0171	102.4 %	0.0179	107.1 %	0.0165	98.7 %
4	0.02198	0.0224	102.0 %	0.0232	105.7 %	0.0216	98.4 %
5	0.02836	0.0287	101.2 %	0.0294	103.8 %	0.0282	99.4 %
6	0.03518	0.0362	103.0 %	0.0372	105.8 %	0.0353	100.3 %
7	0.04084	0.0414	101.3 %	0.0426	104.2 %	0.0399	97.7 %
8	0.04572	0.0470	102.8 %	0.0487	106.5 %	0.0452	98.8 %
9	0.05137	0.0525	102.1 %	0.0544	105.8 %	0.0511	99.4 %
Average Recovery			102.4 %		106.1 %		99.2 %
Standard Deviation			0.86		1.61		0.89

The average recovery of the $BeSO_4 \cdot 4H_2O$ samples on day 0 was 102 % with a range of 101–104 % with a standard deviation of 0.81. On day 1 the average was 106 % with a range of 104–109 % and a standard deviation of 1.61. On day seven the average recovery was 99.2 % with a range of 97.7–100 % and a standard deviation of 0.885.

A summary of the recoveries of beryllium standard from different matrices is listed in Table 3. The average recovery for MCE filters over the range was 97.9 %; the average recovery for PVC filters over the range was 100 %; the average recovery over the range for Whatman® 42 filter paper was 99.2 %; the average recovery over the range for Palintest® wipes was 104 %; and the average recovery over the range for Ghost™ wipes was 100 %. Both the Palintest® and Ghost™ wipes were more variable than the others at the lowest range of 0.001 µg/mL (0.05 µg/sample).

Table 4 details the results of the estimated MDL and MQL investigation. Of the different matrices, Ghost™ wipes had the largest standard deviation and estimated MQL of 0.00023 μg/mL, which corresponds to 0.0058 μg/sample and 0.012 μg/sample for filters and wipes, respectively. Table 5 shows the results of the spiked samples. All matrices gave average recoveries between 94.3 % (MCE filters) and 111 % (Ghost™ wipes). Although the MDL and MQL calculate to lower values, the MQL for all matrices was set at the level of the spiked samples, or 0.0005 μg/mL (0.013 μg/sample for filters and 0.025 μg/sample for wipes). The detection level was set at one-third the MQL for improved accuracy. OSHA previously studied recoveries, the MDL, and the MQL of Ghost™ wipes in method ID-125G [5].

TABLE 3—*Matrix effect of common sampling media.*

Spike (sample concentration μg/mL)	MCE Filter Recovery (a)	PVC Filter Recovery (a)	Whatman® Paper Recovery (b)	Palintest® Wipe Result Recovery (b)	Ghost™ Wipe Recovery (b)
0.001	101.0 %	95.0 %	92.0 %	120.0 %	89.0 %
0.001	98.0 %	97.0 %	92.0 %	141.0 %	94.0 %
0.001	95.0 %	95.0 %	97.0 %	100.0 %	126.0 %
0.01	95.7 %	97.7 %	99.1 %	110.7 %	97.8 %
0.01	95.1 %	98.1 %	99.1 %	101.1 %	96.9 %
0.01	95.4 %	98.0 %	99.5 %	97.6 %	97.0 %
0.1	95.9 %	100.9 %	102.4 %	98.7 %	98.8 %
0.1	97.2 %	103.1 %	100.9 %	99.9 %	99.6 %
0.1	96.5 %	101.7 %	100.2 %	99.6 %	98.0 %
1	98.7 %	99.4 %	102.4 %	99.5 %	98.7 %
1	97.7 %	100.2 %	100.1 %	99.1 %	98.0 %
1	96.1 %	100.3 %	98.3 %	98.8 %	98.7 %
10	98.8 %	105.3 %	99.0 %	100.1 %	101.1 %
10	98.0 %	102.2 %	99.3 %	100.5 %	101.0 %
10	100.1 %	102.8 %	100.4 %	100.0 %	101.5 %
20	102.0 %	103.8 %	101.5 %	99.7 %	102.9 %
20	100.8 %	104.8 %	101.2 %	99.1 %	100.2 %
20	100.5 %	102.0 %	100.4 %	99.3 %	100.1 %
Average	97.9 %	100.4 %	99.2 %	103.6 %	100.0 %
Std. Deviation	0.022	0.031	0.029	0.108	0.072

(a) Final sample volume = 25 mL; (b) Final sample volume = 50 mL.

TABLE 4—*Estimated MDL and MQL for Be.*

Replicate	MCE Filter Result μg/mL (a,b)	PVC Filter Result μg/mL (a,b)	Whatman® Filter Paper Result μg/mL (a,c)	Palintest® Wipe Result μg/mL (a,c)	Ghost™ Wipe Result μg/mL (a,c)
1	0.00002	0.00002	-0.00001	0.00003	-0.00002
2	0.00003	0.00003	0.00003	0.00006	0.00000
3	0.00002	0.00004	0.00002	0.00002	0.00000
4	0.00004	0.00003	0.00003	0.00001	-0.00005
5	-0.00002	0.00005	0.00002	0.00004	0.00001
6	0.00004	0.00004	0.00001	0.00003	-0.00005
7	0.00005	0.00002	0.00002	0.00001	-0.00002
Average	0.00003	0.00003	0.00002	0.00003	-0.00002
Std. Deviation	0.00002	0.00001	0.00001	0.00002	0.00002
EMDL(d)	0.00007	0.00003	0.00004	0.00006	0.00008
EMQL(e)	0.00022	0.00010	0.00013	0.00017	0.00023

(a) Raw instrument result; (b) Final sample volume = 25 mL; (c) Final sample volume = 50 mL (d) EMDL = Estimated MDL; (e) EMQL = Estimated MQL.

TABLE 5—*MQL for Be (liquid spike at 0.0005 μg/mL).*

Replicate	MCE Filter Result μg/mL (a,b)	PVC Filter Result μg/mL (a,b)	Whatman® Filter Paper Result μg/mL (a,c)	Palintest® Wipe Result μg/mL (a,c)	Ghost™ Wipe Result μg/mL (a,c)
1	0.00045	0.00047	0.00048	0.00051	0.00050
2	0.00053	0.00050	0.00054	0.00053	0.00056
3	0.00045	0.00049	0.00048	0.00052	0.00056
4	0.00045	0.00052	0.00051	0.00049	0.00055
5	0.00047	0.00054	0.00049	0.00055	0.00055
6	0.00048	0.00050	0.00054	0.00051	0.00060
7	0.00047	0.00047	0.00050	0.00052	0.00057
Average Recovery	94.3 %	99.7 %	101.1 %	103.7 %	111.1 %
Std. Deviation	0.00003	0.00003	0.00003	0.00002	0.00003
CV	6.1 %	5.1 %	5.1 %	3.6 %	5.4 %

(a) Raw instrument result; (b) Final sample volume = 25 mL; (c) Final sample volume = 50 mL .

Results in Table 6 reaffirm that there is not a digested sample stability problem of BeO and $BeSO_4 \cdot 4H_2O$. There are no initial recoveries listed in Table 6; the experiment was only to check for a possible decreased of recovery over time for Be, BeO, and $BeSO_4 \cdot 4H_2O$.

TABLE 6—*Investigation of sample stability.*

Sample	Day 0 Result (μg/mL)	Day 1 Result (μg/mL)	Day 7 Result (μg/mL)	Std. Deviation	CV
Be-1	5.172	5.239	5.295	0.0617	1.18 %
Be-2	3.828	3.891	3.910	0.0430	1.11 %
Be-3	11.293	11.266	11.646	0.2120	1.86 %
BeO-1	1.723	1.779	2.053	0.1769	9.55 %
BeO-2	2.685	2.710	3.144	0.2581	9.07 %
BeO-3	0.937	0.955	1.056	0.0638	6.49 %
$BeSO_4 \cdot 4H_2O$-1	0.765	0.787	0.761	0.0143	1.85 %
$BeSO_4 \cdot 4H_2O$-2	1.001	1.022	0.989	0.0165	1.65 %
$BeSO_4 \cdot 4H_2O$-3	0.863	0.877	0.851	0.0128	1.49 %

Conclusion

NIOSH NMAM 7303 method can be used to extract beryllium, beryllium oxide, beryllium sulfate, and other soluble beryllium compounds. Although the recovery of beryllium oxide is approximately 80 %, the method may be an alternative to other analytical methods depending on the end user. The advantage of this method is that it can be used to analyze beryllium along with many other metals from a single sample. Also, the MDL and MQL are sufficiently low that the method meets all applicable workplace atmosphere standards for beryllium. Only HCl and HNO_3 are needed for sample preparation, which are commonly used for ICP-AES analysis and do not create viscosity or inter-element interference issues that may be associated with the alternative methods. The method can be used with a variety sampling media including MCE filters, PVC filters, Whatman® 42 90 mm filter paper, Ghost™ wipes, and Palintest® wipes.

References

[1] Rossman, M. D., Pruess, O. P., and Powers, M. B., *Beryllium: Biomedical and Environmental Aspects*, Williams and Wilkins, G. T. Minton, Ed., Baltimore, MD, 1991.
[2] Kolanz, M., "Intoduction to Beryllium: Uses, Regulatory History, and Disease," *App. Occ. and Env. Hyg.*, Vol. 16, No. 5, 2001, pp. 559–567.
[3] *NIOSH Manual of Analytical Methods*, 4th ed., Method No. 7102, U.S. Department of Health and Human Services (NIOSH) Publ. No. 2003-154. NIOSH, Cincinnati, OH, 2003.
[4] *NIOSH Manual of Analytical Methods*, 4th ed., Method No. 7300, U.S. Department of Health and Human Services (NIOSH) Publ. No. 2003-154. NIOSH, Cincinnati, OH, 2003.
[5] Occupational Safety and Health Administration (OSHA), Division of Physical Measurements and Inorganic Analyses, *Metal and Metalloid Particulates in Workplace Atmospheres (ICP Analysis ID-125G*, OSHA Technical Center, Salt Lake City, UT, 1988, 2002 rev.
[6] *NIOSH Manual of Analytical Methods*, 4th ed., Method No. 7303, U.S. Department of Health and Human Services (NIOSH) Publ. No. 2003-154. NIOSH, Cincinnati, OH, 2003.
[7] Wambach, P. F., Tuggle, R. M., "Development of an Eight-Hour Occupational Exposure Limit for Beryllium," *App. Occ. and Env. Hyg.*, Vol. 15, No. 7, 2000, pp. 581–587.

Journal of ASTM International, Jan. 2006, Vol. 3, No. 1
Paper ID: JAI13158
Available online at: www.astm.org

Darryl Campling[1] *and Bharat Patel*[1]

Sampling and Analysis of Beryllium at JET: Policy Cost and Impact

ABSTRACT: This paper describes the sampling and analysis of beryllium at the Joint European Torus (JET) fusion facility. The current policy requiring 100 % personal air sampling (PAS) and taking many surface contamination smears generates 40,000 samples per year. Sample processing, analysis, and the quality assurance (QA) program are described. Costs are summarized to derive a cost per sample of ≈£4.2 ($8). This is economical but as 99.8 % of PAS measurements are $< 2 \mu g/m^3$ and the aggregated costs are high, reviewing the policy is justified. The disadvantages of the present strategy, the options for a change of policy, and the role that an accredited real-time analysis instrument could play are discussed. Retrospective analysis affects the machine operating time; a quicker technique would have a significant impact on improving experimental time. The project's experience is that turnaround times of 2–6 h can be routinely achieved for large numbers of samples.

KEYWORDS: beryllium, fusion, sampling, analysis, policy, cost, real-time

Introduction

The Joint European Torus (JET), located in Oxfordshire, UK, is the world's largest facility for investigating nuclear fusion reactions as a potential future energy source. Since 1989 up to 3 tons of Beryllium (Be) has been installed in the torus as solid components and as an evaporated deposit; its low atomic number, relatively high melting point, and oxygen-gettering properties make it an ideal first wall material [1]. The occupational hygiene aspects of working with Be are controlled by the Health Physics Group who also operate an on-site analysis facility [2].

Legal Status of Be in the UK

Beryllium as a toxic material is controlled under the UK Control of Substances Hazardous to Health Regulations, and has an exposure limit of 2 µg/m3 as an 8 h-TWA Workplace Exposure Limit (WEL) [3]. The WEL label implies that for this substance, exposures be controlled to the lowest level possible, even if measurement shows it to be below the exposure limit.

Health Effects

Even small exposures to Be may produce a physiological effect in some individuals as a hypersensitivity allergic reaction. Such sensitization can then lead to chronic Be disease (CBD), which mainly affects the lungs, causing inflammation and the production of granulomas. Some

---

Manuscript received March 15, 2004; accepted for publication June 29, 2005; published Jan. 2006.
[1] Health Physicist, Euratom/UKAEA Fusion Association, Culham Science Centre, Abingdon, Oxon, OX14 3EA, UK.

forms of Be, notably soluble compounds and Be oxides, are thought to produce a more toxic effect. Long-term handling of Be is also thought to be risk factor for developing lung cancer.

Use of Be at JET

The predominant form of Be used at JET is metal, with much smaller amounts of Be ceramics and alloys. Although most of the handling is of prefabricated metal components, some minor machining has been carried out. Machine operations cause erosion of the first wall surfaces generating some dust. The potential for exposure occurs during entries to the torus vessel or in the breaches of containment when components need to be removed or exchanged. Further exposure can occur in the maintenance and decontamination of components and respiratory protective equipment in Be handling areas. The potential for exposure to higher concentrations is generally limited to the shutdown periods or particular campaigns of work on contaminated items. For most workers, exposure is therefore intermittent rather than continuous, although a few activities, such as waste handling, occur throughout the year.

Health Surveillance

The health surveillance program at JET relies on screening of individuals based on their medical history, making annual lung-function tests, as well as following up reported symptoms of ill health. There is no routine screening of the workforce using the beryllium lymphocyte proliferation test (BeLPT), but such tests could be undertaken as part of a CBD diagnostic evaluation if needed. So far amongst the current workforce there are no reported clinical signs consistent with CBD, although the rate of sensitization (if any cases exist) is not known. The presumption is that, based on this surveillance, the current policy of workplace control and monitoring has been adequate for prevention of CBD.

Be Designated Areas

The project operates 20 permanent Be "controlled" working areas where there is a low-moderate risk of significant Be contamination being encountered if work is taking place [4]. Some of these (eight) are in continuous use with an elevated potential for high levels of surface and airborne, or both, Be contamination. The research nature of the project means that many tasks with Be are not routine and the fluctuations in the use of Be changes the potential for exposures.

There are a further 80 Be "restricted access" areas that are only occasionally entered and where there is only a small chance of low-level Be contamination being encountered. Most of these areas are iso-containers used for the storage of wrapped components. Treating the legacy of waste and used items is liable to generate increased levels in the future.

Current Sampling Policy

The current policy, which has remained in place since 1989, requires the full duration of all entries to all Be controlled areas to be monitored by a personal air sampler (PAS). The wearing of respiratory protective equipment in Be controlled areas is mandatory for all entries to areas where > $0.4 \mu g/m^3$ could be encountered. In practice the vast majority of entries to Be controlled areas utilize some form of respiratory protective equipment even though concentrations are likely to be below 0.4μg/m^3. In addition static air samplers run continuously in all Be controlled areas and are used to survey all tasks where airborne Be contamination might arise. Surface contamination smear surveys are carried out routinely, but with a varying frequency, in all areas. Components and items are required to be surveyed for surface contamination before transfer out of a Be area. An acceptance level of < 10 μg/m^2 (0.1 μg/100 cm^2) is used. This practice is designed to minimize the spread of contamination by ensuring that only clean items are removed from Be areas and that contamination control practices are effective within the areas.

This policy of 100 % PAS and surface clearance sampling was adopted for the following reasons:

- The widespread use of Be in fusion research applications was new and at least initially it was not known what airborne levels might arise in an area or be associated with a particular task.
- The types of maintenance operations were wide ranging, from handling solid machined tiles to grinding or welding in the torus, where mg/m^2 and 10s of μg/m^3 of Be might arise and where pressurized suits would be required for protection.
- It provides reassurance to both the workforce and management that all exposures are monitored. All positive PAS results are subject to investigation.
- The absence of any measurements would require exposure assessments to be made which is time consuming, prone to error, and would prompt a greater level of scrutiny from both staff and the regulator than would otherwise be the case.
- It covers accidental exposures and unexpected situations.
- It prevents non-Be workers from becoming inadvertently exposed by checking surface contamination levels on all transferred items.

Sample Numbers

Approximately 40 000 samples of all types are taken each year, of which between ≈3000 and 14 000 are PAS samples (Fig. 1).

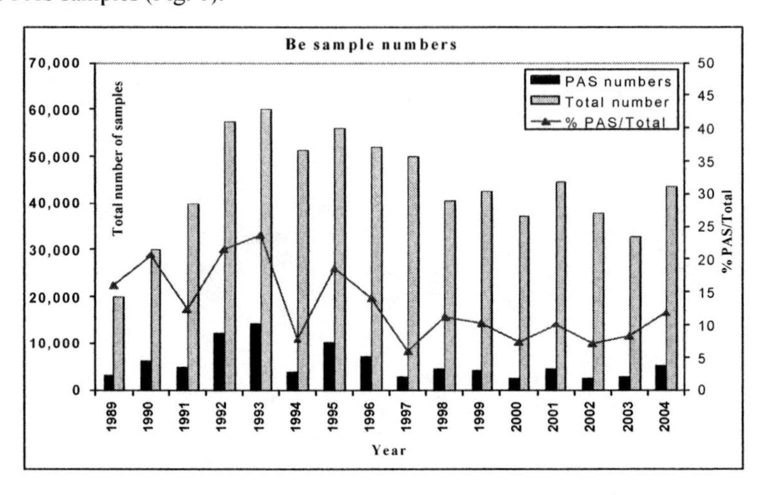

FIG. 1—*Beryllium sample numbers.*

The bulk of the remainder and by far the most numerous are surface contamination smears. Static air samples from work areas and water and oil samples make up a small proportion of the total number of samples. The large variation in the proportion of PAS samples is due to changes in the configuration of the machine, i.e., whether lengthy shutdowns are required and the nature of the maintenance work in each shutdown.

Sample Analysis and Quality Assurance

Surface contamination cellulose filter paper smears and PAS and work area air samples on cellulose-nitrate filter papers are dissolved in 1 ml of 98 % sulphuric acid and 5 ml of 70 % nitric acid at up to 400°C. Automatic sample digestion takes 2 ½ h per batch of 40 samples. The method complies with the UK Health & Safety Executive's specified technique [5]. The end result is a Be-sulphate solution suitable for analysis by flame Atomic Absorption Spectrophotometry (AAS) using a nitrous oxide and acetylene gas mix. Two Perkin Elmer instruments, an 1100B and an Analyst 300, operated manually, achieve an analysis limit of detection (LOD) of 0.03 μg per sample. Sample turnaround times vary between 6 h for routine samples and 1 h for urgent/incident samples, subject to the analysis laboratory being manned. Extended days or shifts are operated if required. Although not currently accredited, the laboratory is pursuing accreditation with the United Kingdom's Accreditation Service (UKAS) to undertake Be analysis.

The quality assurance (QA) surveillance consists of internal and external comparison schemes. For the former, spiked samples are submitted blind to the analyzing technician and the "observed" versus the "expected" values are compared. In the external QA program samples are sent to a contractor's laboratory and analyzed by Induction Coupled Plasma – Optical Emission Spectrophotometry (ICPOES) and again a comparison between the observed and the expected values is made. The results of these checks (carried out since 1990) are summarized in Fig. 2 and show that the differences are just a few percent for all categories of samples.

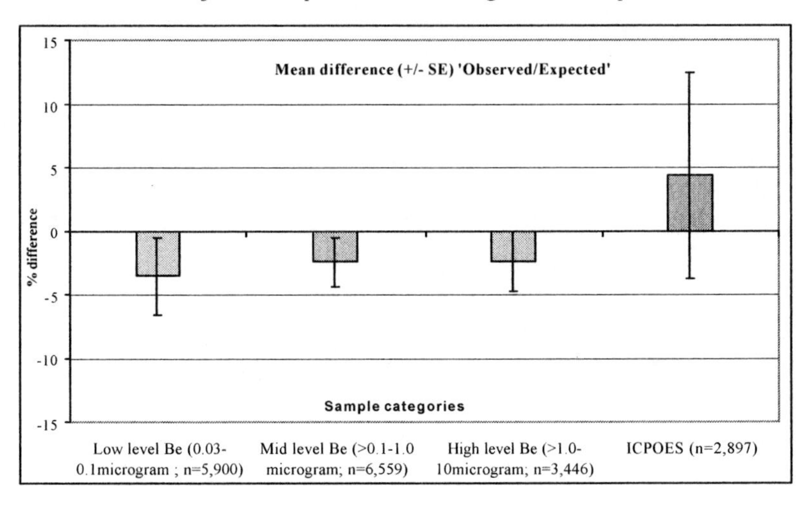

FIG. 2—*Mean difference (± SE) observed/expected quality assurance results.*

The small negative bias on the low, mid, and high level sample categories refers to the in-house QA samples and is due to the small loss of liquor that results when the samples are transferred from the digestion tube or beaker to the calibrated analysis vial. This is considered acceptable as the comparison with the external QA samples carried out by ICPOES shows a similar percentage positive bias.

These values are recorded to facilitate detection of any undesirable trends. New working calibration solutions (1 and 2 μg/ml) are compared to the "in-use" standard and two other standards from different suppliers before they are put into use.

Costs

It costs approximately £37 ($70) million per year to operate JET and the average number of operational days is typically only 150 per year, so maximizing the machine's availability is crucial. Maintenance tasks are often in the torus operations area and the need to set up temporary Be controlled areas and undertake clearance of these areas to restore normal access can result in delays. As such, the operational nature of the project demands the fastest possible sample turnaround times as delays of just a few hours will result in lost operational time. For this reason in 1995 a new purpose-built on-site Be analysis and Health Physics Laboratory was set up at a capital cost of £500k ($960k). The total overall cost including all the major elements required to undertake Be analysis and the running and operational costs over the following 10 years (1995–2004) are summarized in Fig. 3.

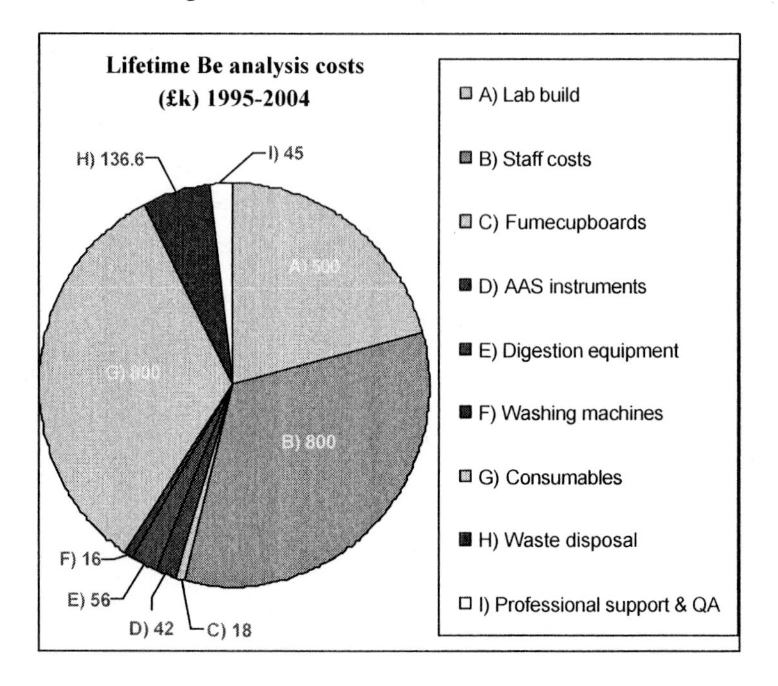

FIG. 3—*Be analysis laboratory lifetime (ten year) running costs.*

Staff numbers consist of, on average, four laboratory technicians, including a supervisor. Three stainless steel fume cupboards were purchased and installed when the laboratory was built; these are expected to last the lifetime of the facility. The cost for AAS instruments includes replacement of obsolete instruments; an older instrument has been replaced. Digestion equipment consists of automatic and manual hotplates of various types and includes replacement of several sets due to wear and tear in a hostile environment. Two industrial duty washing machines have been used for high quality decontamination of glassware, beakers, vials, digestion tubes, etc. The integrated running costs include all the consumables routinely used in the analysis process. A large volume of waste (which is also radioactive) is produced in the analysis process and includes acidic Be contaminated aqueous liquor, soft compactable wastes, contaminated glassware and plastic pipettes, etc. Professional support is required to write risk assessments, produce laboratory standing orders, and oversee quality assurance checks.

The costs associated with air and surface sampling are summarized in Fig. 4. They include the cost of all PAS and static air sample pumps including user-replaceable spare parts and losses due to wear and tear. Also included are the sampling media costs and the manpower costs for the deployment, maintenance, and calibration of the samplers. Professional support takes the form of providing advice as to when sampling will be required and monitoring and recording the PAS results.

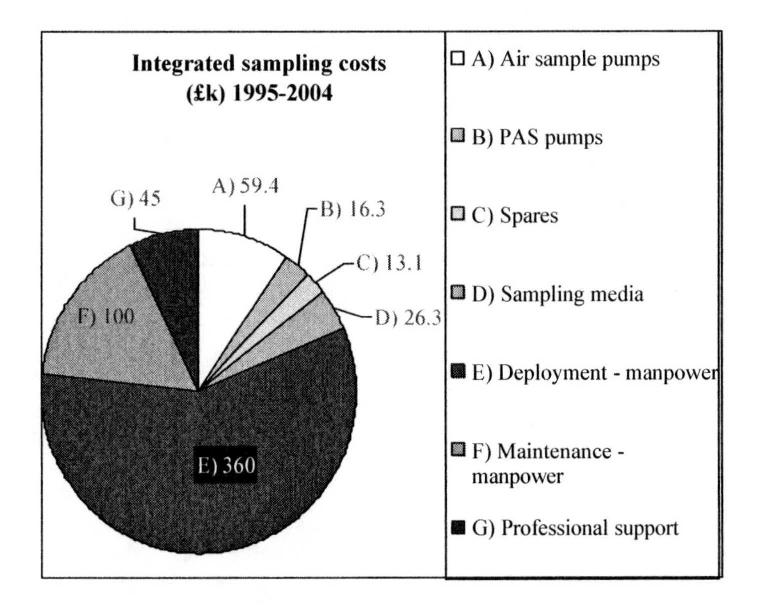

FIG. 4—*Integrated (ten year) air and surface contamination sampling costs.*

The total overall cost for all the elements detailed in Figs. 3 and 4 is £3M ($5.75M). Altogether these costs equate to an average of £4.2 ($8) per sample.

Policy Review

It is clear that the current 100 % PAS sampling policy, the large number of surface contamination smears taken, and the requirement to have fast turnaround times carries a significant cost overhead. The disadvantages of the current policy are the high cost in terms of both manpower and equipment, and the possible diversion of resources from other safety issues.

The original policy was formulated prior to the actual use of Be. Since then considerable experience has been gained, and, on average, the actual exposures have been much lower than anticipated. It is arguable that resources currently devoted to sampling could potentially be more usefully invested in medical screening and surveillance or the improvement of workplace controls.

Allowing for respiratory protective equipment protection factors, 99.98 % of personal exposures are below the UK WEL and a large proportion of surface smears are less than the 10 $\mu g/m^2$ acceptance level. Table 1 summarizes the PAS sampling results from 1989 to the end 2004 and gives the results in seven concentration intervals as a percent of all PAS samples taken. All results above the WEL occurred prior to 1994 when (in the UK) the 2 $\mu g/m^3$ level changed in its legal status from an Occupational Exposure Standard to the present WEL. A similar contamination profile exists for surface smear results.

TABLE 1—*Summary of PAS sampling results, 1989–2004.*

	PAS results. Respiratory protection factor corrected (8-h TWA) Be concentration intervals ($\mu g/m^3$)					
	< 0.03	0.03–0.1	0.1–0.2	0.2–1.0	1.0–2.0	> 2.0
Percent of total (92 332)	94.72	4.83	0.297	0.120	0.018	0.013

In the absence of evidence of direct health effects and a large volume of exposure data it is appropriate to reconsider the extent of the sampling and the validity of the current policy.

A variety of modified policies is feasible and could consist of the following elements:

- Reduce the number of surface contamination smears taken by concentrating on areas and tasks where raised levels are *likely* to be seen. A 60 % reduction might be achieved.
- Retain 100 % PAS but only in areas where raised surface or airborne are known to occur.

Alternatively, a more cautious approach could be:

- Take surface contamination check smears from all areas other than those where contamination is very rarely seen. A 30 % reduction in numbers might be achieved.
- Retain the 100 % PAS sampling policy as it is.

This second set of options recognizes that a change in the WEL is possible in the UK in the next few years and that exposures considered low at present (close to our LOD) may still be of importance [6].

Given the many years of experience, selecting those tasks and areas that could be subject to less rigorous surveillance should be easy. However, this step could be taken more confidently if a prompt assessment of the levels was available with a real-time (or close to real-time) monitor.

Such an instrument would allow selection of just those operations likely to generate significant exposures and would trigger a higher level of surveillance with traditional analysis methods.

Laser Induced Breakdown Spectrometry at JET

In 1989 a prototype Laser Induced Breakdown Spectrometer (LIBS) instrument that could potentially offer much quicker analysis (a few minutes) was tested at JET. Although the instrument worked well with calibration filter papers (produced using a standard solution), a poor correlation was found between twin air sample papers taken from operational areas when one was analyzed by the LIBS instrument and the other by the wet chemistry method. On average the LIBS results were 30 times lower than the AAS results. It is thought that the discrepancy was due to the difference in form between the Be on the calibration papers and that on the work area samples. The Be on the work area papers would have been associated with graphite and in a particulate form. For this reason its use as a survey instrument was not pursued. It should also be noted that use of such an instrument for PAS assessments could only be realized if its performance was validated and the UK's Health and Safety Executive approved it. A suitable monitor would have to reach a sufficiently low LOD, be economic to purchase, easy to operate, and be reliable. The economics of substituting traditional techniques with real-time monitors would have to be considered.

Discussion

Recent epidemiological studies point to the potential for sensitization or CBD in persons exposed to generally low beryllium concentrations [7]. Many of these studies relate to beryllium machining plants and generally to continuous or repeated exposures. While there is much uncertainty about the effects of low level exposures, and further research is needed, there is clearly a need to reduce exposures to levels as low as reasonably practical (ALARP) below the exposure limit. While in the United States there are proposals for reductions in the exposure limit from certain groups representing industrial hygiene standards [8], so far there is no similar indication from the U.S. federal OSHA body, nor from the UK standard-setting authority.

Nevertheless, given that in the UK the statutory Be exposure limit may be reduced within the next few years, it is unlikely that our 100 % PAS policy will be changed. The reassurance of being able to prove that all personal exposures and not just work area levels are ALARP has important legal and worker relations benefits. Even if the current level of PAS and contamination survey sample numbers are not reduced the potential saving of machine downtime would be significant if shorter analysis times could be achieved.

One other major UK facility has reported its results of Be exposure monitoring between 1961 and 1997 [9]. The authors report much higher exposure profiles than those reported here, with one confirmed case of CBD. The processes involved (melting, casting, powder production, pressing, machining, and heat and surface treatments) were much more hazardous than those employed at JET.

Based on our 16 years of experience and much lower exposure levels this gives some reassurance that our strict methods of exposure control may be successful in preventing CBD. It should be pointed out that without routine use of screening tests, such as the Be lymphocyte proliferation test, the rate of sensitization (as opposed to a diagnosis of CBD) is not known in the JET workforce. Therefore there are grounds for retaining both existing workplace controls and the 100 % PAS sampling strategy.

In the field of fusion research the next step machine, the International Thermonuclear Experimental Reactor (ITER), has been designed and is ready to build; only its host country has yet to be agreed upon. As this machine is expected to have a Be first wall and future modifications to JET in 2007–2008 will involve more extensive use of Be, the future benefits on any improvements in Be monitoring are likely to have a significant impact.

The balance of JET experience shows, however, that even with a 100 % sampling policy and large sample numbers, an in-house analysis facility can deliver results within a few hours. Improvements could possibly be made to traditional wet chemistry and AAS techniques by increasing the level of automation.

For an accredited real-time Be analysis instrument to replace traditional mass-based measurement techniques, it must meet a number of operational criteria and in particular show that it is cost effective in saving machine operational time at projects like JET and at any future fusion machines. If future studies indicate that Be disease is caused by sub-micron-sized particles, then a reassessment by us of real-time particle sizing instruments (as compared to analysis instruments) would be required.

Conclusion

Overall the result of this analysis and review is that although there is scope to significantly reduce the number of surface contamination smears, we are unlikely to change the policy with respect to PAS sampling despite the extremely low frequency of positive results. This is on the grounds that the benefits of having a result for each and every exposure provide both workforce and management with a high level of reassurance. In addition, a likely reduction in the maximum exposure limit for Be in the future together with a recognition that the sensitization rate, if any, is unknown indicates that changing this policy now would not be prudent.

With respect to surface contamination smear surveys there is clearly some scope for reducing the number of samples taken. This might save a significant proportion of the analysis laboratories' running costs.

The arrangements for sampling and analysis at JET have largely remained unchanged for over 16 years. Following this review and despite the costs there is strong justification to retain the full scope of sampling. Using traditional analysis methods, sample turnaround times of 2–6 h are routinely achieved and have minimal influence on machine downtime.

References

[1] Patel, B. and Parsons, W., "Operational Be Handling Experience at JET," *Fusion Engineering and Design,* Vol. 69, 2003, pp. 689–694.

[2] Campling, D. C., Litchfield, R. A., and Russ, R. M., "The Operation of the JET Be Analysis Laboratory," *Proceedings of the 18th Symposium on Fusion Technology,* K. Herschbach, W. Maurer, and J. E. Vetter, Eds., Elsevier, 1995.

[3] EH40/2005, Workplace Exposure Limits – Containing the List of Workplace Exposure Limits for Use with the Control of Substances Hazardous to Health Regulations, Her Majesty's Stationary Office, St Clements House, 2–16 Colegate, Norwich NR3 1BQ, ISBN 0717629775, 2002.

[4] Russ, R., "Beryllium safety at JET," *Proceedings of Symposium on Fusion Technology,* 1992.

[5] UK Health and Safety Executive - Method of Determining Hazardous Substances, 29/2 - Be and Be compounds in air (AA), Her Majesty's Stationary Office, St Clements House, 2–16 Colegate, Norwich NR3 1BQ, ISBN 0717611302, 1996.

[6] Kelleher, P. C., Martyny, J. W., and Mroz, M. M., "Beryllium Particulate Exposure and Disease Relations in a Machining Plant," *Journal of Occup. Environ. Med.,* Vol. 43, 2001, pp. 238–249.

[7] Infante, P. F. and Newman L. S., "Beryllium Exposure and Chronic Beryllium Disease," *The Lancet,* Vol. 363, 2004, pp. 415–416.

[8] American Conference of Government Industrial Hygienists, TLVs® and BEL®s, Notice of Intended Changes, 2002 edition.

[9] Johnson, J. S., Foote, K., McClean, M., and Cogbill, G., "Beryllium Exposure Control Program at the Cardiff Atomic Weapons Establishment in the United Kingdom," *Applied Occupational and Environmental Hygiene,* Vol. 16, No. 5, 2001, pp. 619–630.

ON-SITE MONITORING FOR BERYLLIUM—SAMPLING AND ANALYTICAL ASPECTS

Journal of ASTM International, January 2006, Vol. 3, No. 1
Paper ID JAI13172
Available online at www.astm.org

Meng-Dawn Cheng,[1] *Robert W. Smithwick, III,*[1] *and Ray Hinton*[2]

Use of Electrically Enhanced Aerosol Plasma Spectroscopy for Real-Time Characterization of Beryllium Particles

ABSTRACT: The best warning of human exposure to elevated toxic aerosol particles is a monitor that can provide a near-real-time alarm function. Use of surrogate indices such as particle-number concentration, mass concentration, and/or other flow-diagnostics variables is ineffective and could be costly when false positives do arise. We have developed a field-portable system specifically for monitoring beryllium particles in the air in near real-time. The prototype monitor is installed on a two-shelf handcart that can be used in workplaces involving beryllium extraction, machining, and parts fabrication. The measurement involves no sample preparation and generates no analytical waste. The operating principle of the monitor is electrically enhanced laser-induced electrical-plasma spectrometry assisted with aerosol-focusing technology. Performance data of the monitor indicate a dynamic range spanning over four orders of magnitude, and the monitor is capable of detecting an airborne beryllium concentration of 0.05 $\mu g\ m^{-3}$. In reference, the Department of Energy (DOE) standard for beryllium is 0.2 $\mu g\ m^{-3}$ within an 8-h average, while the Occupational Safety and Health Administration standard for beryllium is 2 $\mu g\ m^{-3}$. In addition, the monitor is capable of simultaneous detection of multiple elements using an Echellette spectrometer if needed. The capability of simultaneous detection provides a convenient means for positive identification and possible quantification of multiple elements in near real time. We present the instrument development and calibration data and results from field demonstration conducted at a DOE facility in Oak Ridge, Tennessee.

KEYWORDS: aerosolized beryllium, airborne particles, electrically enhanced laser-based measurement, real-time monitoring

Introduction

Beryllium (Symbol: Be; CAS registry ID 7440-4107, atomic number 4, Group II) is a light element. Beryllium has unique characteristics that make it a superior material for certain specialized applications. Beryllium has a high melting point, a low electrical conductivity, superior strength and stiffness, high thermal conductivity, and high resistance to corrosion. Beryllium is used in several forms: as a pure metal, as beryllium oxide, and as an alloy with copper, aluminum, magnesium, or nickel. Beryllium can be used as X-ray windows (Be transmits X-rays 17 times better than aluminum). A 2 % Be alloy with nickel is used for springs, electrodes and non-sparking tools. Beryllium (2 %) alloyed with copper gives a hard strong alloy with high resistance to wear used in gyroscopes, computer parts, and instruments (desirable lightness, stiffness). Beryllium alloys are also used as a structural material for high-performance aircraft, missiles, spacecraft (such as the USA space shuttle), and communication satellites. It can be used in ceramics, as a moderator in nuclear reactions since it is a highly effective moderator and reflector for neutrons. Beryllium oxide is extensively used in the nuclear industry.

Manuscript received 18 March 2005; accepted for publication 1 August 2005; published January 2006. Presented at ASTM Symposium on Beryllium: Sampling and Analysis on 21-22 April 2005 in Reno, NV.
[1] Oak Ridge National Laboratory, PO Box 2008, MS 6038, Oak Ridge, TN, USA.
[2] Y-12 National Security Complex, PO Box 2009 MS 8189, Oak Ridge, TN, USA.
*Note: Mention of the use of trade names or companies does not imply the endorsement by the authors nor by their associated organizations.

Beryllium has no known, beneficial biological role. In fact, compounds containing beryllium are poisonous. Beryllium metal dust can cause major lung damage and beryllium salts are very toxic. One route for beryllium into the biosphere is by way of industrial smoke. It is well known that exposure to beryllium particles is associated with **c**hronic **b**eryllium **d**isease, CBD [Sterner and Eisenbud (1); Kreiss et al. (2); Kreiss et al., (3); Eisenbud (4)]. CBD is prevalent among beryllium metal, oxides, and copper-alloy workers despite significant reductions in total beryllium particle mass exposure [Eisenbud (4)]. Current occupational exposure limit is 2 μg of beryllium per cubic meter of air (μg m^{-3}) by the Occupational Safety and Health Administration (OSHA), and is 0.2 μg m^{-3} by the Department of Energy (DOE). Those stringent standards have not prevented new cases of CBD [Eisenbud (4)], new investigations have been prompted by the hypothesis that the toxicity of beryllium particles that caused CBD may be related to the size of beryllium particles and the surface area of the particles [Stefaniak et al. (5)].

Measurement of beryllium particles in the air has traditionally been performed by a filter-based collection technique followed by Inductively Coupled Plasma – Atomic Emission Spectroscopy (ICP-AES). This technique is time consuming, labor intensive, prone to contamination, and possibly does not provide fully useful data with which to protect workers' health. It is likely that workers can be exposed to an elevated level of beryllium particles for a short period; thus, the exposure (dose × duration) is still within the OSHA or DOE limit. Such an exposure scenario would not be prevented with the current 8-h Time-Weighted Average (TWA) limit. Also, it is likely that workers might be exposed to low-dose of beryllium in ultrafine or nano-beryllium particles that have low beryllium concentration. Either scenario calls for a more responsive beryllium monitor for use within the workplace. The requirements of the field-portable monitor are listed in Table 1. Some of the requirements are highly demanding and adds serious constraints on the technology to be developed into a field portable platform. The research and development efforts taken by DOE at ORNL and Y-12 facilities have produced such a monitor that meets the requirements and is suitable for use as a real-time alarm within beryllium workplaces.

TABLE 1—*Requirements for a new real-time beryllium monitor.*

No	Description of Requirement
1	Digital output display in μg Be/m^3
2	Minimum detection level 0.05 μg/m^3 (or best available)
3	Measurement range 0.05 to 1,000 μg/m^3
4	Fifteen-min or less response time
5	Measurement accuracy ± 10 % (or best available)
6	Low drift for stable and reliable readings
7	Capability to log/store, and download data to other computer
8	Operation in 0-95 %relative humidity
9	Operation at 32-100°F temperature range
10	Indication that the monitor is working
11	Transportable size for field measurement
12	Corrections for common interferences
13	Correction for carryover from the previous sample

Methods

The real-time Be monitor was developed based on a technique developed by Cheng (6). The original technique utilized a particle-focusing device to concentrate particles for the spectroscopic analysis of elemental composition on particles. Readers interested in further detail of the technique are referred to Cheng (6). We will review related development history and briefly describe the improved technique here. Part of the monitoring system design is based on a time-resolved plasma emission spectroscopic analysis called laser-induced plasma spectroscopy (LIPS) or laser-induced breakdown spectroscopy (LIBS) [Radziemski et al. (7); Neuhauser et al. (8); Essien, Radziemski, and Sneddon (9); Bunkin and Savranskii (10); Yalcin et al., (11); Mokhbat and Hahn (12); Carranza et al. (13); Hahn, Flower, and Hencken (14); Hahn (15); Buckley et al. (16)].

In Aerosol Beam Focused LIPS (ABFLIPS) [Cheng (6)], the laser-induced micro plasma is formed by focusing a high-power laser beam onto the focal point of the particle beam where particles are concentrated by an engineered nozzle (see Fig. 1 for a conceptual drawing). This arrangement facilitates aerosol sampling and plasma formation. When the laser plasma is initiated, the dielectric constant of air is reduced facilitating the formation of second plasma in the confined electrical field. We call this technique an electrically enhanced aerosol plasma spectroscopy (or called EE-APS).

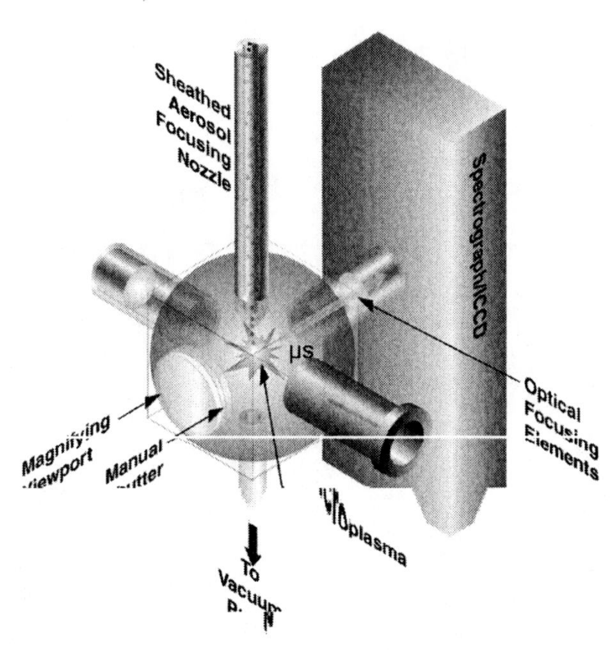

FIG. 1—*Conceptual drawing of aerosol beam focused laser-induced plasma spectroscopy (ABFLIPS).*

The laser-produced spark initiates the electric spark between two electrodes connected to a charged capacitor. For a 3000-volt discharge from a 0.125-μF capacitor, a pulsed energy of about 560 mJ was added to the laser energy of about 90 mJ per pulse. The time between the formation of the first and second spark is within a few nanoseconds. When the plasma cools down between two laser pulses, the emitted light is retrieved remotely, at a 90° angle, by collimating-collection lenses coupled to a 1-m solar-blank fiber-optic cable to a spectrometer equipped with an Echellette spectrograph and an intensified charge-coupled device. The light is analyzed for the spectral lines by custom software running on a Windows®-based PC platform, and the attributes associated with the spectral lines. The area under the singly ionized Be (II) at 313.1-nm peak is calculated and used for quantitative analysis of aerosol beryllium content, while the center wavelength is used for identifying the element. The particle-focusing device we designed serves to introduce the aerosol sample for EE-APS, and to facilitate the formation of the laser-induced micro plasma. The values of instrument parameters are listed in Table 2.

TABLE 2—*Instrumentation parameters.*

Parameters	Description
Spectrograph	Fiber-optic-coupled Echellette spectrograph from Catalina Scientific (Model SE200)
Optical module	UV
Wavelength region	180-900 nm
Focused expanded region	310-316 nm
Blaze offset	-20
Angle offset	1.5
Intensified Charge Coupled Device	Andor iSTAR
Pixel width x height	13 μm x 13 μm
Gate delay	0
Wait time for data collection	4 μs
Pulse width	4 μs
Gain	185
Spectra accumulation time	200s

The concept of using an electrical field to enhance laser-induced plasma signal is not new, but has not been pursued and investigated thoroughly, since it first appeared in open literature [Rasberry et al., (17) and Winstead et al. (18)]. In the EE-APS scheme, a laser pulse initiates the electric discharge with the following advantages: 1) The light emitted from the combined spark can be 100 times to 1000 times more intense than that of the laser spark alone, translating directly into increased sensitivity. 2) the atomic-emission lines are narrower than those of the laser spark alone due to a lower electron density of the electric spark. 3) The broad-background continuum light is lower when the electric spark is used. During early stage of the work, two potential disadvantages were found that the electrode material of the electric spark is a potential interference and that poor precision is often observed, being typically 25-40 % relative standard deviations.

In 1983, researchers at Los Alamos published a paper concerning LIBS of aerosols. In the Introduction, the authors stated that they were aware of electric-spark-assisted LIBS for use with the laser microprobe, but they chose to study the monitoring of beryllium in air with "electrodeless" laser-induced sparks [Radziemski, L.J. et al., (7)]. Since then, beryllium detection by LIPS/LIBS has been primarly based on the electrodeless scheme [e.g., Radziemski, Cremers, and Loree (19); Cremers and Radziemski (20); Ottesen, Wang, and Radziemski (21)].

Our EE-APS technology was tested using beryllium particles produced from nebulizing beryllium-containing solutions. The beryllium standard solution was available from the High Purity Standards, Inc. as a 10.5 μg Be ml^{-1} in 2 % nitric acid (HNO_3) solution. A 6-jet nebulizer (BGI, Inc.) was used to generate liquid droplets that contain the beryllium salt from the standard solution. The produced droplets pass through a thirty-six cm long diffusion-drying column and enter into a 40-L steel drum to be mixed with the particle-free dilution air. The aerosol-generation setup in the laboratory is shown in Fig. 2. The system was designed to allow simultaneous testing of multiple instruments and/or sampling-analytical methods. The sizes of the dried particles follow a lognormal distribution observed by a TSI Scanning Mobility Particle Sizer (SMPS® Model 3080L and Model 3025A). The total number concentration of the particles generated by the nebulizer is generally on the order of 10^5-10^7 cm^{-3} depending on the salt concentration we prepared (based on our prior experience), although we did not monitor the number concentration of the beryllium aerosol at the same time.

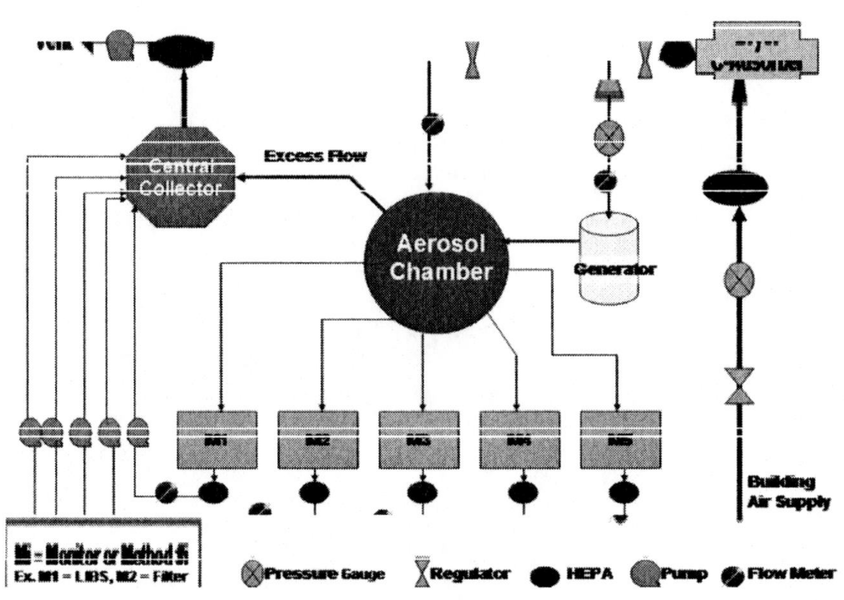

FIG. 2—*Schematic of the beryllium test environment, aerosol generation and conditioning configuration.*

A parallel time-integrated sampling effort was made to collect beryllium particles on TEFLO® filters for subsequent laboratory analysis by ICP-Atomic Emission Spectrometry (ICP-AES). The TEFLO filters have a pore size of 2 μm and its collection efficiency of 0.3 μm Dactyl Phthalate (DOP) particles was rated at greater than 99.99 %. Sizes of particles generated with the nebulizer were in the range of 30 nm to 600 nm measured by the SMPS®. The collection efficiency of the TEFLO filters for the nebulizer-generated particles was greater than 99 % consistent with literature values [Baron and Willeke (2001)]. The result also indicates virtually all airborne particles were captured by the filter, and that it is reasonable to compare the filter values determined by ICP-AES with our instrument readout. The filter analysis was performed by the analytical services laboratory at the DOE/Y-12 National Security Complex using a certified beryllium analytical protocol. The beryllium concentrations (in units of μg Be/m^3 of air) measured by the real-time monitor are then compared to those measured on the collected filters from the experiments. The sampling duration for a filter typically ran from 2 to 4 h depending on the beryllium solution concentration, while our instrument readout of concentrations takes only minutes. The beryllium-in-the-air concentration is calculated by dividing the filtered beryllium mass determined with the ICP-AES by the total air volume sampled.

Subsequent to successful laboratory development and testing, the real-time Be monitor was tested in a beryllium work area at the DOE/Y12 facility. During this field test, the monitor was located in a Be buffer zone (a separate room) 15-m away from the work area. Simultaneous sampling was made by pumping the air through a 0.95-cm inner diameter Tygon tubing at 60 L min^{-1} volumetric flow rate using a vacuum pump located in the buffer zone. A TSI single particle Aerosol Time-Of-Flight Mass Spectrometer (ATOFMS; Model 3800) was co-located and used to measure airborne particles in the buffer zone next to EE-APS. Filter sampling was performed to collect airborne beryllium particles on the TEFLO® filters at a critical-orifice controlled volumetric flow rate of 10 L min^{-1}.

Results and Discussion

Measurement of Laboratory-Generated Beryllium Concentration

Figure 3 shows a comparison between the EE-APS monitoring readings and the filter measurement of beryllium concentration in aerosols. The quantitative signal is taken at 313.1 nm by integrating the area under the peak, while the noise is taken at an off-the-peak wavelength. Every data point shown in Fig. 3 has a signal-to-noise ratio greater than 10. The error bar was estimated using seven replicated measurements, and found to be in the range of ± 12 % to ± 22 %. Note that this is not a full-span calibration curve since we did not test the instrument at high beryllium concentration to obtain the "full dynamic range" of the analytical technique. Our interest has primary been at the developing a technique to obtain a reasonable signal-to-noise ratio at a lower beryllium concentration, particularly at the level of 0.2 μg m^{-3} where the DOE standard dedicates.

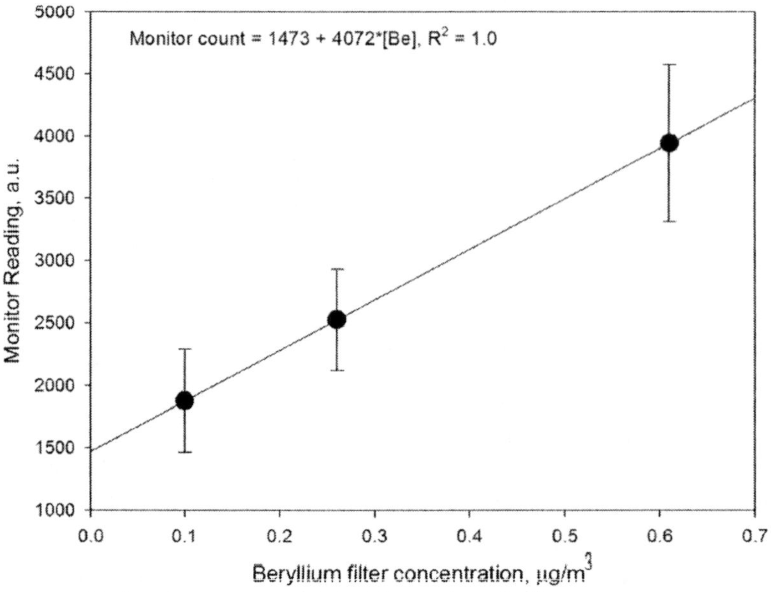

FIG. 3—*Comparison of EE-APS signals with filter-measurements.*

We did not pursue the detection limit for the Be concentration at the present, but the data shown in Fig. 1 suggest the new technique has more than sufficient power to detect the beryllium aerosol at a level below the DOE standard. To demonstrate this can be achieved, we attempted another experiment, this time an aerosol beryllium concentration of 0.05 $\mu g/m^3$ or 50 ng/m^3 was generated. Four examples of the spectral data are presented in Fig. 4. If the average peak height at 313.06 nm (800) is divided by the average off-line background (200), one obtains a signal-to-noise ratio of 4, we conclude that is sufficient (greater than 2) for the quantification purpose. The signals at this low concentration level were also reasonably stable since the fluctuation [as measured by coefficient of variation in % = (standard deviation/average)*100] associated with the Be concentration of 50 ng m^{-3} (measured by the filter method) was approximately 16 %. The fluctuation for blank, particle-free air was approximately 4 %. The sources of difference between 16 % and 4 % can be attributed to combined variation in aerosol focusing, flow stability (pumping), and spark-to-spark variation.

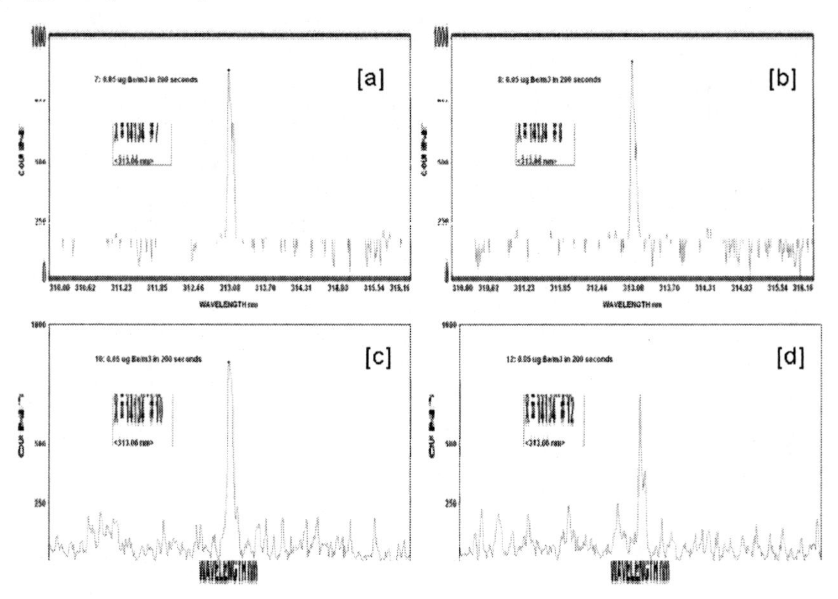

FIG. 4—*Example of four beryllium spectra taken at the airborne beryllium concentration of 50 ng* m^{-3}.

Beryllium Particle Measurement in a Work Place

Aerosol beryllium in a work place at the DOE/Y-12 facility in Oak Ridge was monitored on December 17, 2003. The EE-APS monitor was located in a beryllium buffer area outside of the room. A 15-m long sampling tube was used for transferring potentially beryllium-laden air into the beryllium monitor. The area has no production-like activities performed; clean-up activities were conducted on this day. Some beryllium oxide materials were moved around within a containment box that was located near the end of the sampling tube.

Each spike in Fig. 5 represents a 215-s (3 min and 35 s) measurement interval by EE-APS. Note that the individual concentration of all measurements shown in Fig. 5 is smaller than 0.2

μg/m³ except the sample #12. Measurements # 1 to 12 were estimated by the software. Figure 5 also shows a Be spike immediately after these materials were moved (labeled Run #13), possibly releasing Be-containing aerosols or dusts into the air that was extracted to the EE-APS and the co-located TSI instrument. The TSI aerosol time-of-flight mass spectrometer also detected Be during this same event [Fig. 6*a*]. The TSI instrument also indicated that the beryllium in the aerosol particles that were detected by both instruments could be BeO. The particles detected by the TSI instrument were approximately 3 μm in size obtained from the time-of-flight data. The TSI instrument is a single-particle instrument; while EE-APS measures ensemble average of Be concentration of several particles. No quantitative data about the beryllium concentration were provided by ATOF-MS. This data set was used to corroborate the EE-APS identification only. Other elements were also identified, not quantified, by the ATOF-MS. Runs #14-16 appear to be recovering from the Be excursion observed in Run #13. Another spike was observed during Run #21 which corresponded to the time in which a beryllium worker (who was in full protective clothing and respirator) wiped off the ends of the sampling tubes prior to moving the sampling tubes to another location. Figure 6*b* shows the ATOF-MS identification of Be particles for this event.

The TSI instrument and EE-APS could not see the same particles even though they are sampled from one source. For a realistic comparison is to use laboratory-generated particles of known properties. Even under that condition, the sample intervals of the two instruments still need to be compatible to yield comparable results. The ATOF-MS spectra shown in two panels in Fig. 6 were from two single particles, and the EE-APS results for the same intervals shown in Fig. 5 were from more than two particles. Other particles in the population could give off different ATOF-MS spectra. In short, both instruments positively identified Be in aerosol at the work place, providing encouragement that EE-APS can be used as a sensitive real-time Be monitor.

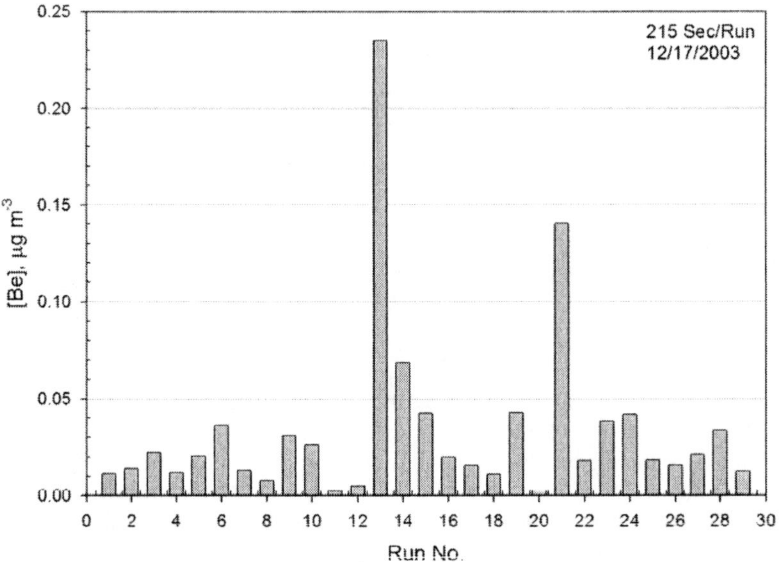

FIG. 5—*Beryllium concentration* ($\mu g\ m^{-3}$) *measured by EE-APS during the test at the Y12 beryllium work area.*

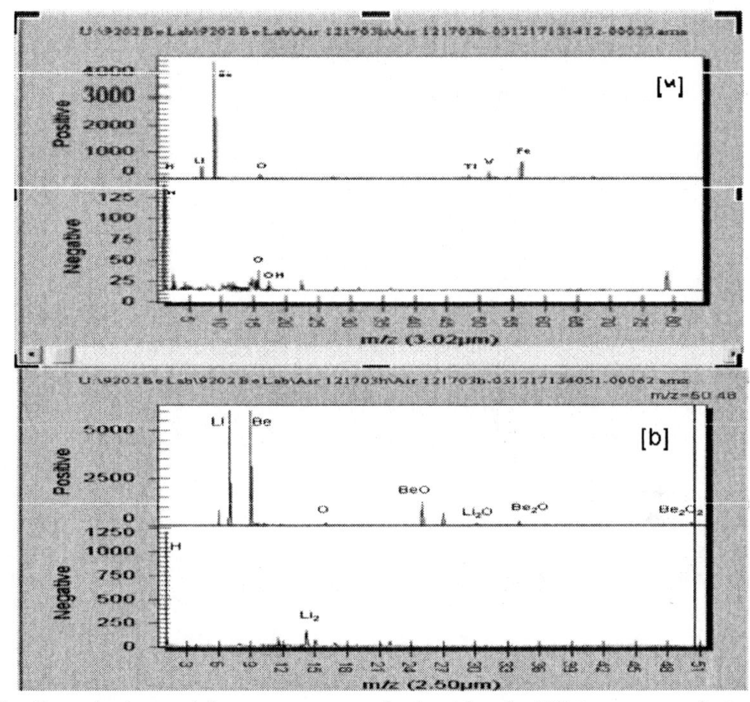

FIG. 6—*Two single-particle mass spectra obtained by the TSI instrument during the times EE-APS observed the big spikes.*

Conclusions

An instrument has been developed for a DOE/Y-12 facility for real-time monitoring of Be in the air. During the field demonstration, the instrument was located 15 m away from a beryllium work area, and has shown promising results. In the laboratory evaluation during the development, the instrument showed a high sensitivity in detecting Be particles in the air at mass concentration of 50 ng m^{-3} in about three minutes. Such real-time measurement ability for beryllium particles in the air has not been reported previously in the open literature. With the combination of our patented particle-beam focusing device, laser-induced plasma spectroscopy, and our new patent-pending electrical enhancement, we have achieved an unprecedented sensitivity for beryllium measurement. This ability could be further developed by the commercial sector into a real-time alarm/monitor for protection of the health of beryllium workers. Finally, the technique we reported here is applicable to other metals embedded in aerosol particles. Our technique does not distinguish particulate metals from the gas if exist at all, the technique measures total content of metals in the sample. The field measurements by our technique were positively correlated with a commercial time-of-flight single particle instrument.

Recommendations

The information provided by this technique is real time and can prevent worker's exposure to highly toxic material, such as beryllium particles, even if the exposure is transient. A small, portable monitor can be built based on the technique described here, and it can be used to monitor and evaluate different operational procedures involving beryllium particles. The monitor can also be used as an alarm during decontamination and demolition activities that involves beryllium materials, and possibly other toxic metals.

Acknowledgments

The authors acknowledged the financial supports provided by the Plant Directed Research, Development, and Demonstration Program Office of the Department of Energy/Y-12 National Security Complex. Robert Steele is the program manager. MDC also acknowledges the supports of the Department of Defense Environmental Security Technology Certification Program. The cooperation and technical support of Debbie Davidson, Vic Upchurch, Craig Hanzelka, Tom Oatts, and Carl Lyster are acknowledged. The authors acknowledged Raymond Hinton for his management support and Stephen Cristy for sharing the single particle time-of-flight mass spectroscopy data and laboratory facilities. The Y12 analytical chemistry organization provided timely and professional support for the beryllium filter analysis and data certification. Oak Ridge National Laboratory is managed by UT-Battelle, LLC, for the U.S. Dept. of Energy under contract DE-AC05-00OR22725. Y-12 National Security Complex is managed by BWXT Y-12, L.L.C. for the U.S. Department of Energy Under Contract DE-AC05-00OR22800.

References

[1] Sterner, J. H. and M. Eisenbud (1951) Epidemiology of Beryllium Intoxication. A. M. A. Arch. Ind. Hyg. Occup. Med. 4(2):123-151.

[2] Kreiss, K., M. M. Mroz, L. S. Newman, J. W. Martyny, and B. Zhen (1996) Machining Risk of Beryllium Disease and Sensitization with Median Exposure below 2 $\mu g\ m^{-3}$. *Am. J. Ind. Med.* 30:16-25.

[3] Kreiss, K., M. M. Morz, B. Zhen, H. Wiedemann, and B. Barna (1997) Risk of Beryllium Disease Related to Work Processes at a Metal, Alloy, and Oxide Production Plant. *Occup. Environ. Med.*, 54:605-612.

[4] Eisenbud, M. (1998) The Standard for Control of Chronic Beryllium Disease. *Appl. Occup. Environ. Hyg*, 13:25-31.

[5] Stefaniak, A. B., M. D. Hoover, R. M. Dickerson, E. J. Peterson, G. A. Day, P. N. Breysse, M. S. Kent, and R. C. Scripsick (2003) Surface Area of Respirable Beryllium Metal, Oxide, and Copper Alloy Aerosols and Implications for Assessment of Exposure Risk of Chronic Beryllium Disease, *AIHA J.*, 64:297-306.

[6] Cheng, M.-D. (2003) Field Measurement Comparison of Aerosol Metals Using Aerosol Beam Focused Laser-Induced Plasma Spectrometer and Reference Methods, *Talanta*, 61:127-137.

[7] Radziemski, L. J., T. R. Loree, D. A. Cremers, and N. M. Hoffman (1983) Time-Resolved Laser-Induced Breakdown Spectrometry of Aerosols, *Anal. Chem.*, 55:1246-1252.

[8] Neuhauser, R. E., U. Panne, R. Niessner, and P. Wilbring (1999) On-Line Monitoring of Chromium Aerosols in Industrial Exhaust Streams by Laser-Induced Plasma Spectroscopy (LIPS), *Fresenius J. Anal. Chem.*, 364:720-726.

[9] Essien, M., L. J. Radziemski, and J. Sneddon (1988) Detection of Cadmium, Lead and Zinc in Aerosols by Laser-Induced Breakdown Spectrometry, *J. Anal. Atomic Spectrometry*, 3:985-988.

[10] Bunkin and Savranskii (1974) Optical Breakdown of Gases Induced by the Thermal Explosion of Suspended Macroscopic Particles, *Sov. Phys. JETP*, 38:1091-1096.

[11] Yalcin, S., D. R. Crosley, G. P. Smith, and G. W. Faris (1996) Spectroscopic Characterization of Laser-Produced Plasmas for In Situ Toxic Metal Monitoring, *Hazardous Waste & Hazardous Materials*, 13:51-61.

[12] Mokhbat, E. A. and D. W. Hahn (2002) Laser-Induced Breakdown Spectroscopy for the Analysis of Cobalt-Chromium Orthopaedic Wear Debris Particles, Appl. Spectrosc., 56:984-993.

[13] Carranza, J. E., B. T. Fisher, G. D. Yoder, and D. W. Hahn (2001) On-Line Analysis of Ambient Air Aerosols Using Laser-Induced Breakdown Spectroscopy, *Spectrochimica Acta Part B* 56:851-864.

[14] Hahn, D. W., W. L. Flower, and K. R. Hencken (1997) Discrete Particle Detection and Metals Emissions Monitoring Using Laser-Induced Breakdown Spectroscopy, *Appl. Spectrosc.*, 51:1836-1844.

[15] Hahn (1998) Laser-Induced Breakdown Spectroscopy for Sizing and Elemental Analysis of Discrete Aerosol Particles, *Appl. Phys. Ltr.*, 72:2960-2962.

[16] Buckley, S. G., H. A. Johnsen, K. R. Hencken, and D. W. Hahn (2000) Implementation of Laser-Induced Breakdown Spectroscopy as a Continuous Emission Monitor for Toxic Metals, *Waste Manage.*, 20:455-462.

[17] Rasberry, S. D., B. F. Scribner, and M. Margoshes (1967) Laser Probe Excitation in Spectrochemical Analysis. I: Characteristics of the Source, *Appl. Opt.*, 6:81-93.

[18] Winstead, C. B., K. X. He, D. Grantier, T. Hammond, and J. L. Gole (1991) Electric-Field-Enhanced Laser-Induced Plasma Spectroscopy of Jet-Cooled Silicon Trimer, *Chem. Phys. Ltr.*, 181: 222-231.

[19] Radziemski, L. J., D. A. Cremers, and T. R. Loree (1983) Detection of Beryllium by Laser-Induced-Breakdown Spectroscopy, *Spectrochimica Acta Part B*, 38:349-355.

[20] Cremers, D. A. and L. J. Radziemski (1985) Direct Detection of Beryllium on Filters Using the Laser Spark, *Appl. Spectrosc.*, 39:57-63.

[21] Ottesen, D. K., J. C. F. Wang, and L. J. Radziemski (1989) Real-Time Laser Spark Spectroscopy of Particulates in Combustion Environments, *Appl. Spectrosc.*, 43:967-976.

[22] Baron, P. A. and Willeke, K. eds. (2001) Aerosol Measurement: Principles, Techniques, and Applications, 2nd Edition, Wiley, New York, NY.

Journal of ASTM International, October 2005, Vol. 2, No. 9
Paper ID JAI13168
Available online at www.astm.org

Edel M. Minogue,[1] *Deborah S. Ehler,*[2] *Anthony K. Burrell,*[3]* *T. Mark McCleskey,*[3*] *and Tammy P. Taylor*[4]

Development of a New Fluorescence Method for the Detection of Beryllium on Surfaces

---

ABSTRACT: A rapid, quantitative, sensitive test for the detection of beryllium on surfaces has been developed. The method is based on the fluorescence of beryllium bound to sulfonated hydroxybenzoquinoline at pH 12.8, which emits at 475 nm when excited at 380 nm and includes a novel dissolution technique. The intensity of fluorescence is linear with respect to beryllium concentration. A detection limit of 0.02 μg Be/100 cm^2 has been achieved, which is ten times lower than the DOE recommended working limit for non-beryllium work areas [1]. Interference studies have been carried out with a variety of metals including Al, Fe, Pb, U, Cd, Cr, Hg, Ca, W, Ni, Co, and Cu with minimal or no interferences found for detection of Be at 100 nM in the presence 0.4 mM of the other metal. The method has proven successful under various operating conditions, including the detection of beryllium on a variety of surfaces both in laboratory settings and in field trials. It fulfills the requirements for a fast, inexpensive, field deployable method of detection of beryllium on surfaces.

KEYWORDS: beryllium, fluorimetric detection, HBQS, environmental monitoring

Introduction

The unique properties of beryllium (Be) have led to many applications ranging from the aerospace and nuclear industry to manufacturing and electronics. Unfortunately, beryllium is a Class A EPA carcinogen and when inhaled into the lungs can cause the incurable and potentially fatal lung disease, chronic beryllium disease (CBD). Therefore, the monitoring of beryllium in occupational environments is of vital importance. Congress has recently passed limits of Be exposure of 2 $\mu g/m^3$ [2], and DOE facilities have adopted even more stringent levels that include 0.2 $\mu g/m^3$ for airborne levels and 0.2 μg/100 cm^2 as a surface level for the release of items from beryllium areas [1]. For release to another DOE facility working with beryllium, contamination levels are not to exceed 3.0 μg/100 cm^2 [1].

To date, the standard method for the detection of beryllium on surfaces is a surface swipe technique described by OSHA (ID-125G) [3]. The method involves swiping a 10 cm × 10 cm area with a cellulose ester membrane and subsequently digesting the membrane with hydrogen

---

Manuscript received 16 February 2005; accepted for publication 31 March 2005; published October 2005.
Presented at ASTM Symposium on Beryllium: Sampling and Analysis on 21-22 April 2005 in Reno, NV.
[1] Postdoctoral Research Associate, Chemistry Division (C-SIC), Los Alamos National Laboratory, MS J514, Los Alamos, New Mexico 87545, USA.
[2] Chemical Research Technician, Chemistry Division (C-SIC), Los Alamos National Laboratory, MS J514, Los Alamos, New Mexico 87545, USA.
[3] Technical Staff Member, Chemistry Division (C-SIC), Los Alamos National Laboratory, MS J514, Los Alamos, New Mexico 87545, USA.
[4] Technical Staff Member, Nuclear Nonproliferation (N-4), Los Alamos National Laboratory, MS E541, Los Alamos, New Mexico 87545, USA.
* Correspondence may be addressed to either author: E-mail (A.K.B.): burrell@lanl.gov, (T.M.M.) tmark@lanl.gov.

peroxide and sulfuric acid. Inductively coupled plasma atomic emission spectroscopy (ICP-AES) is used to quantify beryllium in the samples. Although straightforward, the procedure can be costly, turnaround time is slow, and it is unsuitable for field use. In addition, the current OSHA method requires consumption of the entire sample in order to meet detection levels. Consequently, verification of results can be difficult if concerns arise post-analysis. There have been attempts to develop a swipe analysis technique based on absorbance changes [4,5], but they have been unable to obtain the necessary quantitative detection limits of 0.02 μg/100 cm^2 for NIOSH approval.

Fluorescence is an ideal method of detection because it is extremely sensitive, is non-destructive, and can be performed quickly. Fluorescent detection of Be has been reported since the 1950s with literature reports on a variety of fluorescent indicators including morin [6,7,8], chromotropic acid [9], and Schiff bases [10]. Despite the many reports of fluorescent indicators for Be, a complete system for the fluorescence detection of Be has yet to be approved by NIOSH, and there is no commercial fluorescent Be detector kit. A complete, robust fluorescent detection method requires three key features: a dissolution method that is able to dissolve Be and BeO and remains compatible with the fluorescence indicator; tolerance to a wide variety of interferences; and a minimal number of simple steps from dissolution to detection. Typical dissolution methods for the dissolution of BeO from a swipe involve concentrated inorganic acid and heating; in addition some methods use hydrogen peroxide. Such conditions are not compatible with any known fluorescent indicator, so the solution must be evaporated to dryness and further treated before it can be added to the fluorescent indicator. The work presented herein is a description of the development of a rapid fluorescence method for the quantitative detection of beryllium on surfaces using the indicator 10-hydroxybenzo[h]quinoline-7-sulfonate (10-HBQS). The method is beryllium specific, inexpensive, applicable to different swipe materials, and field deployable. Detection limits of 0.02 μg beryllium/100 cm^2 swiped surface (one tenth of the DOE required action level of 0.2 μg/100 cm^2) have been achieved. We are currently working with the National Institute for Occupational Safety and Health (NIOSH) for approval of this method for beryllium detection.

In order to eliminate the time-consuming and non-fieldable digestion steps of current standard methods, the use of a fluoride-based medium to dissolve Be was investigated. It was found that Be metal was dissolved within seconds in 1 % ammonium bifluoride $(NH_4)HF_2$. However, high-fired BeO is the most difficult form of Be to dissolve. We tested the dissolution of 10 mg quantities of BeO with 50 mL of 1 % $(NH_4)HF_2$ to demonstrate that 80 % of the oxide form could be dissolved in just 15 min with minimal agitation. Fluoride, usually in the form of HF, is well noted for its ability to penetrate and dissolve metal oxides [11]. Most fluorescent indicators reported do not tolerate the presence of fluoride. The few reports of indicators that tolerate fluoride have complicated procedures involving heating with acid for dissolution and a titration process to obtain the final pH. The duration and complexity of those procedures do not lend themselves easily to field analysis.

Having screened several potential ligands, 10-HBQS, a water-soluble fluorescent dye, was selected for the development of the fluorescence method. The selection of 10-HBQS stemmed from work done in a previous study by Matsumiya et al. [12], where they studied beryllium in urban air and showed that 10-HBQS, hydroxybenzoquinoline (HBQ) chelated the Be(II) ion. In another work, they used the precursor HBQ as a pre-column chelating reagent for the determination of beryllium in water by reversed-phased high-performance liquid chromatography [13]. HBQ fluorescent detection involves the formation of a six-membered chelate ring with Be. A tightly bound hydrogen bonded proton leads to weak triplet emission at

580 nm. When the proton is displaced by a metal such as beryllium, fluorescence emission is observed at 475 nm. However, because HBQ is sparingly soluble in water, we selected the sulfonated derivative 10-HBQS for our studies. Although HBQ was previously commercially available, neither HBQ nor 10-HBQS are currently commercially available, nor are there useful synthetic procedures published. Therefore, we developed synthetic pathways for both of these compounds [14].

Experimental

Apparatus

A miniature fluorescence spectrometer from Ocean Optics (S2000-FL) was customized to incorporate a UV LED with an excitation wavelength of 380 nm (continuous mode). Instrument calibration was carried out using a LS-1-CAL white light source. Detection was carried out using the USB2000 Miniature Fiber Optic Spectrometer connected to the serial port of a laptop computer. Spectra were obtained in the relative irradiance mode using Ocean Optics OOIBase32 Software. The wavelength of emission is 475 nm. The detection limit of the set-up was 0.06 ppb Be. Results were verified by ICP-AES, Jobin Yvon Inc., Edison, New Jersey. This particular instrument has a detection limit of approximately 100 ppt Be, allowing good comparison with the low levels of detection obtainable with our fluorimetric method. Verification of side-by-side swipes was carried out by the standard method for detection of beryllium on surfaces (i.e., digestion of swipe and then ICP-AES). The pH was measured using an Orion pH/ISE Model 710 meter, which was calibrated using pH 4, 7, and 10 buffer solutions (Fisher Scientific Inc.).

Reagents and Solutions

Solid forms of beryllium used included beryllium oxide (BeO 99 %, Acros) and beryllium sulfate ($BeSO_4$, Acros). All solid forms of beryllium were handled in a HEPA-filtered glove box by a beryllium-trained worker. The following stock solutions were prepared: $(NH_4)HF_2$ (Aldrich), 1 % wt/vol in water, 1.1 mM HBQS pH adjusted to pH 12 with 10 M NaOH (Fisher), 100 mM L-Lysine monohydrochloride (Aldrich) at pH 11-12, and 1 mM EDTA disodium dihydrate (J.T. Baker, Inc.). ICP standard solutions (1000 μg/mL metal; SPEX Centriprep) of the following metals were used in interference studies: Al, U, Ca, Li, Pb, Zn, Fe, V, Sn, W, Cu, Ni, Co, Cd, Cr, and Hg. Deionized water (MilliQ®) was used throughout.

Whatman® 541 filters (47 mm diameter) are used as the standard swipe in our experiment. These cellulosic filters are currently utilized by Los Alamos National Laboratory industrial hygienists for the NIOSH approved method of Be testing and from this point will be referred to as *swipes*. The term *filter* will be used when a surface has not been swiped (e.g., for experiments where filters are spiked with known concentrations of Be).

Method

General Procedure

The detection reagent was prepared by the addition of 12.5 mL of 10.7 mM EDTA and 25 mL of 107 mM L-lysine monohydrochloride to 3 mL of 1.1 mM 10-HBQS. The pH was adjusted to 12.85 with the careful addition of 10 M NaOH and water added to a total of 50 mL. Beryllium standards were generated using Be spectrometric standard solutions diluted into 1 % $(NH_4)HF_2$

for the desired concentrations. For calibration curves a 0.1-mL aliquot of each standard solution was added to 1.9 mL of the detection reagent, and spectra were taken at a set integration time. A linear increase in intensity at 475 nm with respect to increasing beryllium concentration was observed (Fig. 1). This enabled the conversion of intensities to concentrations. The amount of Be (μg/100 cm^2) in the area swiped (A) was then obtained by Eq 1, whereby C_s (μg/L) is the concentration for a given sample with a volume of V_s (L), and C_b (μg / L) is the concentration of the blank with a volume of V_b (L). F_d is the dilution factor in this method:

$$[Be](\mu g/100cm^2) = \frac{F_d \times [C_s V_s - C_b V_b]}{A} \tag{1}$$

A result of 2 ppb in our method corresponds to 0.2 μg Be on the swipe. Results must be normalized if an area greater than 100 cm^2 is swiped. If the concentration of beryllium is out of range (too high), then the instrument is recalibrated using higher standards and a shorter integration time. In this way, the range of analysis can be extended. For quality control purposes, a calibration standard and a reagent blank are analyzed at least once every 20 samples.

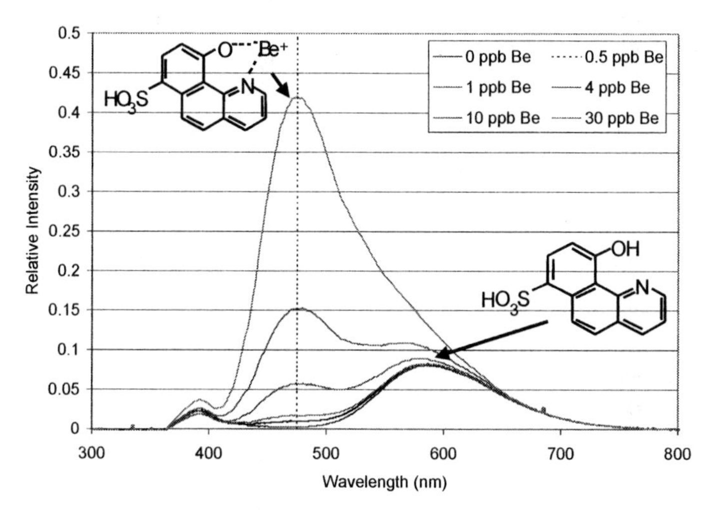

FIG. 1—*Characteristic spectra for HBQS bound (475 nm) and unbound (580 nm) to Be.*

Dissolution Study

The dissolution study was comprised of two areas of interest: the suitability of the Be-dissolving agent and the time-minimization of this step. Preliminary studies of dissolution show that 1 % $(NH_4)HF_2$ dissolves Be and BeO at levels within the required detection range (i.e., 0.02 μg–3.0 μg Be/swipe). Moreover, $(NH_4)HF_2$ does not interfere with 10-HBQS, the ligand of choice. Time analyses were carried out in order to minimize the dissolution time while ensuring that beryllium was dissolved. A 0.15 μg BeO/mL suspension was made by adding 7.5 μg of BeO to 50 mL H_2O. A filter was spiked with a 5-μL aliquot of the suspension. The spiked filter was placed in a tube, and 5 mL of 1 % $(NH_4)HF_2$ was added, the tube capped and then rotated. A 0.5-mL aliquot was taken at set intervals and added to 1.5 mL of the dye reagent mix in a cuvette. Spectra were taken for each interval, and the intensity at 475 nm observed. A series of 10 filters was spiked with the BeO suspension, analyzed by the fluorescence procedure, and then compared to ICP results by measuring the filtrate and the filter by ICP with microwave digestion to ensure all BeO was dissolved.

Interference Study

The following metal solutions were made by dissolving the standard ICP metal solution with 1 % $(NH_4)HF_2$ such that the end concentration of the 0.1 mL aliquot in the 1.9 mL dye mix was between 0.04 mM and 2.0 mM: 0.4 mM Al, 0.4 mM U, 2.0 mM Ca, 0.4 mM Li, 0.4 mM Pb, 0.4 mM Zn, 0.4 mM Fe, 0.4 mM V, 0.4 mM Sn, 0.4 mM W, 0.4 mM Cu, 0.4 mM Ni, 0.4 mM Co, 0.04 mM Cd, 0.04 mM Cr, 0.04 mM Hg. Each sample was prepared in triplicate with (100 nM and 1 μM Be) and without Be. The interference metals were in $\geq$50 000-fold molar excess to the Be present. Spectra were taken for each sample, and the intensity at 475 nm was observed.

Stability Study

Both the stability of the detection reagent solution and the Be-$(NH_4)HF_2$ detection reagent solution were studied over time. A 100-mL solution of the detection reagent containing 10-HBQS, EDTA and buffer was made as previously described. 1.9-ml aliquots were removed at set time intervals, and 0.1 ml of Be standards in $(NH_4)HF_2$ were added and analyzed fluorimetrically. The stability of the final samples was tested by keeping the first set of standards sealed in cuvettes, which were subsequently fluorimetrically analyzed on a weekly basis.

Detection Limit

The current required NIOSH detection limit is 0.2 μg Be/100cm^2. In order to quantify the method detection limit, the following standards were prepared: five low-level standards (0.02 μg - ten times lower than the required detection limit), five standards at the detection limit of 0.2 μg, one standard of 0.1 μg, and a reagent blank. Filters were spiked with the standards and dried for 20 min, after which time 5 mL of $(NH_4)HF_2$ was added, followed by fluorimetric analysis.

Procedure for the Swipe Test

A 100-cm^2 surface was swiped with a Whatman® 541 filter moistened with deionized water, in accordance with the procedure described in OSHA ID-125G [3] and in ASTM D 6966 [15]. The swipe was then placed into a 15-mL polypropylene tube, and 5 mL of the 1 % -$(NH_4)HF_2$ solution was added. The tube was capped and then rotated (Barnstead/Labquake tube rotator) for 30 min, during which time the Be was dissolved. The solution was filtered through a luer-locked PTFE (Millipore) or nylon 0.45-μm syringe filter. In a disposable, clear-sided cuvette, 0.1 mL of the filtrate was added to 1.9 mL of the dye solution mix (20× dilution). The cuvette was capped and briefly shaken, and a fluorescence spectrum was taken ($\lambda_{excitation}$ = 380 nm; $\lambda_{emission}$ = 475 nm). A set of Be standards using the same dye mix was also prepared, and the fluorescence spectra were taken for each set of samples. A calibration curve of the intensities of Be at 475 nm versus beryllium concentration was plotted. From this, the beryllium concentration in the sample was obtained. The remaining Be filtrate was analyzed using ICP-AES, providing corroborative results.

Field Trials

The implementation of our fluorimetric method on swipes from different environments was investigated. Potentially, Be-contaminated surfaces were swiped according to OSHA and NIOSH procedures [3] by an industrial hygienist at Los Alamos National Laboratory in the laboratory, in the beryllium workshop areas, and also in the field. A 100-cm^2 area was swiped and the swipe placed in a tube. A 5-mL aliquot of $(NH_4)HF_2$ was added to the tube, which was subsequently rotated for 30 min. The Be-$(NH_4)HF_2$ solution was decanted into a luer-locked syringe filter and filtered. A 0.1-mL aliquot of the filtrate was added to 1.9 mL of the detection reagent, and the sample was fluorimetrically tested for Be. The remaining filtrate was sent to ICP-AES for confirmational results.

In addition to this, 100 μL of potential interferents such as ethylene glycol, oil, and cleaning agents, were added to Be-spiked filters. The filters were then subject to fluorimetric analysis. This was carried out in duplicate.

Side-by-side swipes from both a Be contaminated shop and firing points including surfaces such as steel, aluminum, and paint were also collected, with one swipe analyzed by the fluorimetric method and the other by the digestion/ICP-AES method. The remainder of the Be-$(NH_4)HF_2$ filtrate was also analyzed by ICP-AES.

Results and Discussion

Fluoride Interference with Indicator

Based on preliminary experiments involving the dissolution of BeO with $(NH_4)HF_2$, we needed a fluorescent indicator that could tolerate large concentrations of fluoride. HBQS had previously been reported to tolerate up to 20 000 000 equivalents of fluoride [12]. Most other Be fluorescent indicators are readily susceptible to fluoride interference at only 300 equivalents. We tested the response of HBQS in the presence of 0.25 % fluoride and found that it responded well. The increase of intensity at 475 nm with respect to beryllium concentration as exhibited in Fig. 1 is not only a indication of the effectiveness of the ligand 10-HBQS, but also is proof of the effectiveness of the ligand in the matrix containing $(NH_4)HF_2$.

Dissolution Study

The dissolution of Be from the swipe into the $(NH_4)HF_2$ solution is the time-limiting step for this otherwise instantaneous method. We minimized this by investigating the time dependence for the dissolution of high fired BeO, one of the most inert forms of Be, spiked onto a Whatman® 541 filter. The BeO used in this study was obtained from Aldrich and has been fired at 2000°C. The intensity of the sample at 475 nm increased with increasing dissolution time up until 25 min. A direct overlap of the intensities at 25 min and 30 min was observed. No further increase of the fluorescence was observed. Therefore, 30 min was chosen as the dissolution time for our experiments, providing a quick response time and near-complete dissolution. Studies comparing the fluorescence technique to ICP measurements on the same solution showed >83 % recovery of BeO in all cases. A consistent amount of residual solution is left on the filter, but there was no evidence of un-dissolved BeO remaining on the filter.

Interference Study

Interference studies with a range of other metals have shown that even in 50 000-fold molar excess over Be, metals such as Pb, U, Hg, or Cr show little (<1 %) or no interference (Table 1). The exception was that high concentrations of Fe (i.e., >20 μM Fe) have a negative effect on Be intensity of approximately 10 % because suspended Fe precipitate absorbs light at 380 nm. If, however, the Fe precipitate is allowed to settle for 4 h or is filtered using a PTFE or nylon filter, and is then reanalyzed, there is no interference. Having the Fe precipitate is an advantage of working at a high pH. Therefore, it is recommended that, with fluorimetric analysis of beryllium, if high iron content is suspected (e.g., due to swiping a rusty surface) or is evident from the gold-orange color that appears when the HBQS mix is added, filter the solution or allow the solution to settle until clear and colorless, and then carry out the fluorimetric analysis.

TABLE 1—*Interference study.*

	Relative Intensity at 475 nm		
	0 Be	100 nM Be	1 μM Be
No Interferents	0.005	0.112	1.078
0.4 mM Al	0.004	0.112	1.054
0.4 mM U	0.004	0.110	1.060
2.0 mM Ca	0.004	0.112	1.057
0.04 mM Li	0.004	0.112	1.060
0.4 mM Pb	0.004	0.111	1.105
0.4 mM Zn	0.003	0.112	1.103
0.4 mM Fe	0.003	0.101	0.925
0.4 mM V	0.003	0.114	1.083
0.4 mM Sn	0.003	0.113	1.105
0.4 mM W	0.003	0.116	1.103
0.4 mM Cu	0.003	0.114	1.062
0.4 mM Ni	0.004	0.114	1.074
0.4 mM Co	0.005	0.111	1.030

Stability Study

For the development of a field deployable method, it is essential that the reagents are stable over a given period of time. Therefore, the stability of the dye mix solution (stored in brown Nalgene HDPE bottles) was studied over time by running Be calibration curves made with the aging dye. After 120 days, no decrease in response was observed. Beryllium standard solutions, which contained the dye mix solution, were also studied over time. They remained stable over 28 days, thus enabling rapid on-site detection of beryllium with pre-prepared reagents and standards. It should be noted that if the beryllium standards including the dye mix are to be stored for longer than a week, the solutions should be stored in a screw-topped, sealable container.

Detection Limit

The method limit of detection (LOD) and the instrument detection limit were determined according to NIOSH procedures [16]. The low-level calibration standards were analyzed and the average result obtained for replicate aliquots. The results obtained were graphed against the mass of Be, and the linear regression equation $Y = mX + c$ enabled the evaluation of responses, Y^*_i, for Be mass. The standard error of regression was calculated using Eq 2, where N is the number of data points, Y^*_i is the predicted value from the least squares fit, and Y_i is the experimental value:

$$s_y = \left[\frac{\Sigma\left(Y_i^{*} - Y_i\right)^2}{(N-2)} \right]^{\frac{1}{2}} \tag{2}$$

A limit of detection of 13.6 ng / swipe (0.136 ppb) was achieved from Eq 3 below:

$$LOD = \left(\frac{3s_y}{m} \right) \tag{3}$$

Field Trial of Swipe Test

The Be-$(NH_4)HF_2$ solutions from field swipes were analyzed by both the fluorimetric method and ICP-AES. The recovery rate was 99.5 %, reinforcing the suitability of the method to realistic environments (Table 2). Beryllium levels ranged from below the fluorimetric detection limit <0.02 μg to 10 μg per area swiped, which far exceeds the threshold limit of 0.2 μg Be/100 cm^2. All were detectable using the method developed and were in concurrence with the results obtained from ICP-AES. No interference was detected when possible contaminants were added to Be-spiked filters. In fact, a 100 % Be recovery rate was observed from filters contaminated with lubricating oil, cutting fluid, and certain cleaners, the exception being Fantastic® spray cleaner for which a 96 % Be recovery rate was observed. A comparison of results from side-by-side swipe analysis highlights the accuracy of this method when compared with the ICP-AES method (Fig. 2). It is difficult to compare side-by-side swipes, as they are not actually swiping the exact same area, but these results indicate that the fluorimetric method can stand up to even the toughest test. Neither method showed consistently higher or lower biased values.

TABLE 2—*Beryllium recovery analysis from samples taken in field trials.*

	Be (μg /100 cm^2)			
Sample No.	Filtrate*	Filtrate#	Residual on Swipe#	% Recovery
2003-01923				
A	0.347	0.350	ND	100
B	0.137	0.130	ND	100
C	0.134	0.120	ND	100
D	0.002	0.020	ND	100
E	5.950	6.150	0.048	99.20
F	5.425	5.400	0.052	99.05
G	5.143	5.500	0.035	99.32
H	3.179	3.210	0.047	98.54
I	6.423	6.600	0.192	97.10
J	2.182	2.030	0.034	98.46
K	4.236	4.170	0.099	97.72
L	1.137	1.050	ND	ND
M	0.007	0.020	ND	ND

* Measured by fluorimetric method; # Measured using ICP-AES; ND: Not detected.

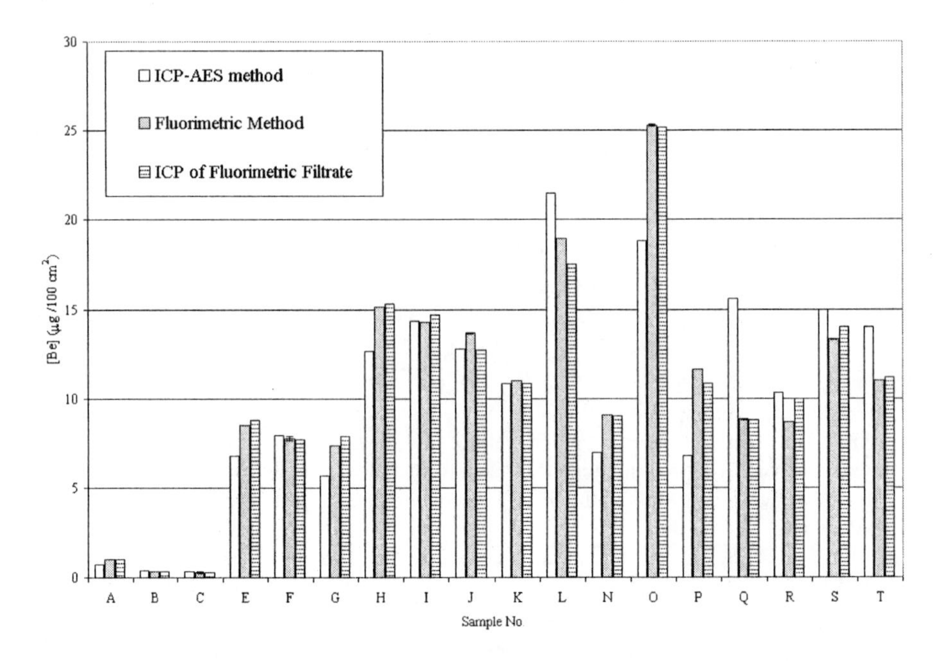

FIG. 2—*Comparison of results obtained from side-by-side swipes including the comparison of the results obtained by the fluorimetric method and the ICP analysis of the fluorimetric solution.*

Conclusions

The method developed is currently undergoing approval by NIOSH as the standard method for the detection of beryllium on surfaces. It is a rapid technique in which beryllium can be extracted from a swipe, bound to a fluorescent dye, and analyzed by fluorescence. We have developed the first complete system for Be detection that dissolves both Be and BeO, detects Be down to 0.02 μg/swipe, tolerates a wide range of interferences, and is simple to use. Our method involves placing a swipe in a dissolution solution, mixing for 30 min, transferring a small aliquot to a detection solution, and measuring the fluorescence at 475 nm. The potential portability of the fluorimetric device coupled with simplicity and specificity of the chemistry lends itself well to field analysis. Work is currently being carried out on the integration of this method into a portable sensor platform.

Acknowledgments

The authors acknowledge the help of B. Duran (C-SIC) for ICP-AES analysis and G. Whitney (HSR-5) and C. Brink (C-ACS) for field trials.

References

[1] CFR (Code of Federal Regulations), Title 10, Energy, Part 850, Department of Energy, 2001 parts 500 to end.
[2] Kolanz, M. E., *Appl. Occupational and Environ. Hygiene*, 2001a, Vol. 16, No. 5, pp. 559–567.
[3] "Metal and Metalloid Particulates in Workplace Atmospheres," (ICP Analysis), Method No. ID-125G, Control No. T-ID-125G-FV-03-0209-M, http://www.osha.gov/dts/sltc/methods/inorganic/id125g/id125g.html#table1.
[4] Taylor, T. P. and Sauer, N. N., "Beryllium Colorimetric Detection for High Speed Monitoring of Laboratory Environments," *J. Hazard. Mater.*, B93, 2002, pp. 271–283.
[5] Taylor, T. P., Ding, M., Ehler, D. S., Foreman, T. M., Kaszuba, J. P., and Sauer, N. N., "Beryllium in the Environment: A Review," *J. Environ. Sci. Health*, A38, 2, 2003, pp. 439–469.
[6] Laitinen, H. A. and Kivalo P., "Fluorometric Determination of Traces of Beryllium," *Anal. Chem.*, Vol. 249, No. 1, 1952, pp. 467–1471.
[7] Capitan, F., Manzano, E., Navalon, A., Vilchez, J. L., and Capitan-Vallvey, L. F, "Determination of Beryllium in Water by Ion-Exchange Spectro-Fluorimetry," *Analyst*, Vol., 114, No. 8, 1989, pp. 969–973.
[8] Saari, L. A. and Seitz, W. R., "Optical Sensors for Beryllium Based on Immobilized Morin Fluorescence," *Analyst*, Vol. 109, No. 5, 1984, pp. 655–657.
[9] Pal, B. K. and Baksi, K., "Chromotropic Acid as a Fluorogenic Agent. 1. Fluorometric Determination of Beryllium," *Microchim Acta*, 1992, Vol. 108, pp. 275–283.
[10] Morisige, K., "Metal-Complexes of Aromatic Schiff-base Compounds. 1. Fluorescence Properties of Aluminum and Gallium Complexes of Aromatic Schiff-bases and their Use in Fluorimetry," *Anal. Chim. Acta*, Vol. 73, No. 2, 1974, pp. 245–254.
[11] Cotton, F. A. and Wilkinson, G., *Advanced Inorganic Chemistry*, 5th ed., Wiley-Interscience, New York, 1988.
[12] Matsumiya, H., Hoshino, H., and Yotsuyanagi, T., "A Novel Fluorescence Reagent, 10-hydroxybenzo[h]quinoline-7-sulfonate, for Selective Determination of Beryllium(II) Ion at pg cm(-3) Levels," *Analyst*, Vol. 126, 2001, pp. 2082–2086.
[13] Matsumiya, H. and Hoshino, H., "Selective Determination of Beryllium(II) Ion at Picomole per Decimeter Cubed Levels by Kinetic Differentiation Mode Reversed-Phase High-Performance Liquid Chromatography with Fluorometric Detection Using 2-(2 '-hydroxyphenyl)-10-hydroxybenzo[h]quinoline as Precolumn Chelating Reagent," *Anal. Chem.*, Vol. 75, 2003, pp. 413–419.
[14] Collis, G. E. and Burrell, A. K., *Tetrahedron Lett*, submitted.
[15] ASTM Standard D 6966, "Standard Practice for Collection of Surface Wipe Samples for Subsequent Determination of Metals," Annual Book of ASTM Standards, ASTM International, West Conshohocken, PA, 2003.
[16] Kennedy, E. R., Fischbach, T. J., Song, R., Eller, P. M., and Shulman, S. A., "Guidelines for Air Sampling and Analytical Method Development and Evaluation," DHHS (NIOSH), Publication No. 95-117, 1995.

Journal of ASTM International, October 2005, Vol. 2, No. 9
Paper ID JAI13156
Available online at www.astm.org

Kevin Ashley,[1] *T. Mark McCleskey,*[2] *Michael J. Brisson,*[3] *Gordon Goodyear,*[4] *John Cronin,*[4] *and Anoop Agrawal*[4]

Interlaboratory Evaluation of a Portable Fluorescence Method for the Measurement of Trace Beryllium in the Workplace*

ABSTRACT: Researchers at Los Alamos National Laboratory (LANL) developed a field-portable fluorescence method for the measurement of trace beryllium in workplace samples such as surface dust and air filters. The technology has been privately licensed and is commercially available. In cooperation with the Analytical Subcommittee of the Beryllium Health and Safety Committee, we have carried out a collaborative interlaboratory evaluation of the LANL field-portable fluorescence method. The interlaboratory study was conducted for the purpose of providing performance data that can be used to support standard methods. Mixed cellulose ester (MCE) membrane filters and Whatman 541 filters were spiked with beryllium standard solutions so that the filters spanned the range ≈0.05 – ≈0.5 μg Be per sample. Sets of these filters were then coded (to ensure blind analysis) and sent to participating laboratories, where they were analyzed. Analysis consisted of the following steps: 1. Removal of the filters from transport cassettes and placement of them into 15-mL centrifuge tubes; 2. mechanically-assisted extraction of the filters in 5 mL of 1 % ammonium bifluoride solution (aqueous) for 30 min; 3.-4. filtration and transfer of sample extract aliquots (100 μL) into fluorescence cuvettes; 5. introduction of 1.9 mL of detection solution (to effect reaction of the fluorescence reagent with beryllium in the extracted sample); and 6. measurement of fluorescence at ≈475 nm using a portable fluorometer. This work presents performance data in support of a procedure that is targeted for publication as a National Institute for Occupational Safety and Health (NIOSH) method and as an ASTM International standard.

KEYWORDS: beryllium, field-portable, fluorescence, interlaboratory evaluation, on-site monitoring, trace analysis, workplace

Introduction

Occupational exposure to beryllium can cause insidious and sometimes fatal disease, and new exposure limits for beryllium in air and on surfaces have been established in efforts to reduce exposure risks to potentially affected workers [1]. Advances in sampling and analytical methods for beryllium are needed in order to meet the challenges relating to exposure assessment and risk reduction. Accurate knowledge of the level of beryllium metal present in the workplace environment is crucial for the determination of the health risks posed to workers.

Field-portable techniques for the accurate, expeditious, and cost-effective monitoring of beryllium are desired to enable rapid assessment of potential worker exposures to this toxic metal

[1] U.S. Department of Health and Human Services, Centers for Disease Control and Prevention, National Institute for Occupational Safety and Health, 4676 Columbia Parkway, Mail Stop R-7, Cincinnati, OH 45226-1998 (USA); tel. +1(513)841-4402; fax +1(513)841-4500; e-mail: KAshley@cdc.gov.
[2] Los Alamos National Laboratory, P.O. Box 1663, MS J-582, Los Alamos, NM 87545, USA.
[3] Westinghouse Savannah River Company, Savannah River Site 707-F, Aiken, SC 29808, USA.
[4] Berylliant, Inc., 4541 E. Fort Lowell Road, Tucson, AZ 85712, USA.

in the occupational environment. These considerations have resulted in efforts to develop field-portable analytical methods for measuring trace concentrations of beryllium on-site in the workplace. Candidate techniques for beryllium field monitoring have included fluorescence [2] and electroanalysis [3].

In the last few years, a field-portable fluorometric method was developed by researchers at Los Alamos National Laboratory (LANL) [4]; this method has recently been licensed and marketed commercially [5]. Owing primarily to the use of a novel fluorophore for Be^{2+} ion [6], hydroxybenzoquinoline sulfonate, the LANL field method offers significantly better limits of detection (LODs) for beryllium than were attainable by using fluorometric reagents investigated earlier. The previous methods [7,8] relied on fluorescence reagents that demonstrate insufficient sensitivity for trace measurements of beryllium, which are now required in workplace settings. More recent investigations have proposed new fluorometric techniques using reagents that enable ultratrace beryllium measurement in the laboratory [4,9].

The objective of the present study was to carry out an interlaboratory evaluation of the on-site fluorometric method for beryllium as it is currently marketed. An aim of this work was to establish estimates of method performance based on a collaborative interlaboratory analysis. These method performance parameters can then be used to support governmental methods such as those published by the National Institute for Occupational Safety and Health (NIOSH) [10]. Also, it is intended that method performance data obtained through this interlaboratory trial will be used as a basis for voluntary consensus standards, such as those published by ASTM International [11].

Performance Evaluation Samples

Performance evaluation material samples (PEMs) consisted of beryllium (in solution and diluted from standard beryllium nitrate solutions using deionized water) pipetted onto mixed-cellulose ester (MCE) membrane filters (Millipore, Billerica, MA) and Whatman® 541 cellulose fiber filters (SKC, Inc., Eighty-Four, PA). The filters were fortified at known levels between ≈0.05 and ≈0.5 μg Be per sample; the volume of the spiking aliquot was 0.1 mL. Also included were blanks of each sample medium ("spiked" with pure deionized water). After spiking by using micropipettes, the spiked filters were then allowed to dry in air at ambient temperature.

To ensure consistency with an ASTM International standard practice pertaining to interlaboratory testing [12], PEMs consisting of blanks plus sampling media spiked at four loading levels (0.050, 0.10, 0.20, and 0.40 μg Be per sample) were prepared. These PEMs were prepared with beryllium loadings targeted to bracket new action levels of 0.2 μg per 100-cm^2 sampling area for surface wipe samples [1] and 0.2 μg m^{-3} for 8-h time-weighted average (TWA) air filter samples [13]. The PEMs were prepared at a contract laboratory (Environmental Resource Associates, Arvada, CO; Lot no. 0809-04-04) under the oversight of LANL. PEMs were subsequently repackaged by the CDC/NIOSH Quality Assurance Coordinator to ensure blind analyses by the participating laboratories.

Interlaboratory Evaluation

Participating laboratories consisted of a subset of prospective participants that were identified by members of the Analytical Subcommittee of the Beryllium Health and Safety Committee [14]. PEMs were mailed to each volunteering laboratory by the coordinating laboratory

(CDC/NIOSH, Cincinnati, OH). Each participating laboratory, along with associated PEM samples, was assigned a numerical code in order to ensure anonymity.

It was requested that the participating laboratories prepare and analyze the PEMs in accordance with the marketed procedure and kit [5]. Briefly, the analysis procedure consisted of the following steps (schematized in Fig. 1):

1. Removal of the filter samples from transport cassettes and placement of them into 15-mL plastic centrifuge tubes
2. Mechanically-assisted extraction of the filters in 5 mL of 1 % ammonium bifluoride solution (aqueous) for 30 min (in 15-mL centrifuge tubes mounted in a mechanical shaker)
3. Filtration of the extracted solutions through plastic syringe microfilters
4. Transfer of sample extract aliquots (100 μL) into fluorescence cuvettes using mechanical pipettes
5. Introduction of 1.9 mL of fluorescent dye detection solution to effect reaction of the fluorescence reagent with beryllium in the extracted sample
6. Measurement of fluorescence at ~475 nm using a portable fluorometer

The participating laboratories were asked to report analysis results in units of mass of beryllium (in μg) per PEM sample. (This required comparison of results for unknowns with calibration standards, along with consideration of appropriate dilution and correction factors, to convert fluorescence intensity to mass [5].)

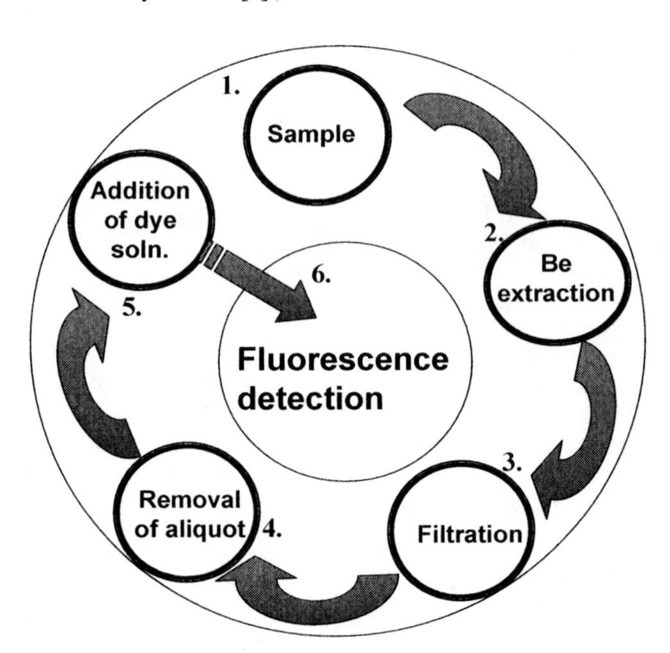

FIG. 1—*Scheme for field-based preparation and fluorescence analysis of workplace samples for determination of beryllium content.*

Precision, Bias, and Statistical Analysis

ASTM International voluntary consensus standard test methods require estimates of measurement uncertainty, and this can be in the form of precision and bias data [15]. Precision estimates are preferably obtained through data from interlaboratory evaluations. Bias of a test procedure must be estimated by evaluating the performance of the test method in question against a reference method, or from determinations of recoveries from the analysis of reference material samples, or both.

In this investigation, interlaboratory precision of analytical results from the volunteer laboratories was examined using statistics from overall interlaboratory analysis results. For purposes of satisfying ASTM International standard criteria, the analysis was done using statistics described in ASTM Standard E 691 [12], which requires a minimum of six participating laboratories. This standard practice also recommends a minimum of four samples for each type of matrix, with duplicate analyses of each of the four samples. Thus, each laboratory received a total of ten PEM samples (five for each filter matrix) for analysis by the field-portable fluorescence method for beryllium. An analogous interlaboratory validation study has been carried out previously in order to evaluate field-portable electroanalytical procedures for on-site determination of lead in environmental samples [16].

Repeatability and reproducibility were calculated for each of the four beryllium levels in the PEMs analyzed by the participating laboratories. Repeatability is an estimate of within-laboratory variability, while reproducibility is an estimate of the variability of both within- and between-laboratory results. Repeatability was calculated by averaging the squares of the standard deviations of within-laboratory results for each beryllium level; hence the average within-laboratory variance is given by the repeatability variance, $(S_r)^2$. Reproducibility variance is expressed by:

$$(S_R)^2 = (S_r)^2 + (S_L)^2$$

where S_L is the sample standard deviation of the mean value estimated from the average of reported interlaboratory test results for a given PEM. Relative standard deviations (RSDs) for repeatability and reproducibility (RSD_r and RSD_R, respectively) are then computed by dividing the standard deviations S_r and S_R by the mean interlaboratory test result for a particular PEM. The RSDs calculated can then be compared with the minimum precision that is desired (e.g., RSD=0.20 [17]) for the test method under evaluation.

Estimates of analytical bias, B, were computed by simply dividing the difference between the measurand and the reference value by the reference value:

$$B = (\mu_i - R_i) / R_i$$

Here, μ_i and R_i are the mean and reference beryllium contents, respectively, for the ith beryllium loading level in each PEM sample.

Results of the Interlaboratory Evaluation

Of the candidate volunteer laboratories that were identified by members of the Analytical Subcommittee of the Beryllium Health and Safety Committee, PEM filter samples were sent to eleven prospective laboratories. Analysis results were subsequently reported from eight laboratories, thereby exceeding the minimum number (for evaluation of an ASTM International test method) of six participants. Hence, for purposes of this round-robin evaluation, recruitment of a sufficient number of volunteers was achieved.

Analytical results reported by the eight individual laboratories that participated in the interlaboratory exercise are summarized in Table 1 for MCE filters and in Table 2 for Whatman 541 filters. For six of the eight laboratories, duplicate analyses were reported for PEM samples at each beryllium loading level for different sample aliquots run using (a) different portable fluorescence spectrometers, or (b) different fluorescence intensity integration times, or (c) both. Overall means were computed based on the pooled means for the average of the two results reported by each laboratory for each sample (excepting the two laboratories that reported a single result). Data from blank measurements were all near to or below the reported LOD of the analytical method (≈0.01 μg Be per sample) [5].

Results for repeatability and reproducibility for the two PEM filter matrices, determined in accordance with ASTM E 691 [12], are summarized in Table 3. Bias estimates for each PEM sample containing beryllium are presented in Table 4; overall mean values μ_i used in estimations of bias were taken from Tables 1 and 2 (for MCE and Whatman filter PEM samples, respectively).

TABLE 1—*Results from measurement of beryllium content in MCE filters, as reported by laboratories participating in the interlaboratory evaluation. Reference values for beryllium loadings on the PEM filter samples are given in parentheses in the column headings.*

Laboratory Number (n = 8)	Low (0.05 μg Be)	Medium Low (0.10 μg Be)	Medium High (0.20 μg Be)	High (0.40 μg Be)
001	0.0512; 0.0504	0.104; 0.105	0.203; 0.207	0.468; 0.484
003	0.060; 0.050	0.11; 0.10	0.21; 0.20	0.43; 0.40
005	0.052; 0.063	0.103; 0.125	0.222; 0.273	0.459; 0.503
006	0.050	0.10	0.21	0.41
007	0.0505; 0.0490	0.103; 0.103	0.210; 0.198	0.406; 0.396
009	0.051; 0.041	0.103; 0.092	0.208; 0.199	0.421; 0.421
010	0.053; 0.053	0.104; 0.107	0.197; 0.194	0.412; 0.415
011	0.053	0.105	0.203	0.404
Overall mean ± sample standard deviation	0.052 ± 0.0038	0.10 ± 0.0048	0.21 ± 0.016	0.43 ± 0.032
Relative Standard deviation	0.073	0.048	0.076	0.074

TABLE 2—*Results from measurement of beryllium content in Whatman 541 filters, as reported by laboratories participating in the interlaboratory evaluation. Reference values for beryllium loadings on the PEM filter samples are given in parentheses in the column headings.*

Laboratory Number (n = 8)	Low (0.05 μg Be)	Medium Low (0.10 μg Be)	Medium High (0.20 μg Be)	High (0.40 μg Be)
001	0.0528; 0.0519	0.103; 0.104	0.198; 0.203	0.399; 0.406
003	0.060; 0.050	0.11; 0.10	0.22; 0.20	0.42; 0.40
005	0.055; 0.063	0.114; 0.145	0.198; 0.251	0.437; 0.492
006	0.050	0.11	0.21	0.40
007	0.0530; 0.0500	0.103; 0.099	0.203; 0.198	0.401; 0.393
009	0.056; 0.052	0.101; 0.093	0.207; 0.198	0.409; 0.410
010	0.053; 0.057	0.106; 0.104	0.205; 0.209	0.403; 0.412
011	0.056	0.104	0.207	0.409
Overall mean ± sample standard deviation	0.054 ± 0.0028	0.11 ± 0.0099	0.21 ± 0.0078	0.41 ± 0.022
Relative Standard Deviation	0.052	0.090	0.037	0.054

TABLE 3—*Repeatability and reproducibility for beryllium measurements from performance evaluation MCE and Whatman 541 filters, as computed using values reported by laboratories (n = 8) participating in the interlaboratory evaluation.*

Beryllium Level	Average (μg Be)	S_r	S_R	RSD_r	RSD_R
MCE Filters					
Low	0.052	0.0034	0.0051	0.065	0.098
Medium Low	0.10	0.0052	0.0071	0.052	0.071
Medium High	0.21	0.012	0.020	0.057	0.095
High	0.43	0.0080	0.033	0.019	0.077
Whatman 541 Filters					
Low	0.054	0.0027	0.0039	0.050	0.072
Medium Low	0.11	0.0068	0.012	0.062	0.11
Medium High	0.21	0.012	0.014	0.057	0.067
High	0.41	0.012	0.025	0.029	0.061

TABLE 4—*Bias estimates for beryllium measurements from performance evaluation MCE and Whatman 541 filters, computed using mean values from Tables 1–3. Reference values for beryllium loadings on the filters are given in parentheses.*

PEM Matrix	Low (0.05 μg Be)	Medium Low (0.10 μg Be)	Medium High (0.20 μg Be)	High (0.40 μg Be)
MCE filters	0.040	0.0	0.050	0.075
Whatman 541 filters	0.080	0.10	0.050	0.025

Discussion

Results shown in Tables 1 and 2 give estimates of interlaboratory precision (as measured by the relative standard deviation, RSD) that are similar for both MCE and Whatman 541 filters. For each loading level there are no statistically significant differences between the mean beryllium contents measured in the two different sampling media (*t*-tests for independent means; n = 8). Also, for both media (Tables 1–3), there is no apparent trend of precision changing as a function of beryllium loading. It is noted that no outlier tests were conducted on the data which were reported by the participating laboratories; all results were included and treated statistically, despite the possibility of statistical outliers. The highest intralaboratory RSD encountered is 0.065, and all interlaboratory RSDs are 0.11 or less (Tables 1–3). Ordinarily, interlaboratory precision estimates of 0.15 and below are regarded as acceptable for PEMs such as these, that is, consisting of liquid spikes on sampling media [16].

The results summarized in Table 3 show that figures for within-laboratory precision RSD_r spanned the range ≈0.02 – ≈0.07, while data for between-laboratory precision RSD_R were slightly greater, ranging from ≈0.06 – ≈0.11. These precision estimates compare very favorably with precision estimates from interlaboratory results for PEMs consisting of MCE filters spiked with beryllium in liquid form at similar levels (Beryllium Proficiency Analytical Testing [BePAT] program, American Industrial Hygiene Association [AIHA], 2003) [18]. For AIHA BePAT PEM samples, interlaboratory RSDs of ≈0.06 – ≈0.15 (n = 25) were computed from five different loading levels ranging from ≈0.15 – ≈0.6 μg Be per filter. The AIHA BePAT samples were prepared and analyzed by laboratories using reference analytical methods involving concentrated acid digestion and atomic spectrometric analysis, for example NIOSH Method 7102 [19]. Thus, it is shown that, for filter samples, the interlaboratory precision of the field-portable fluorescence method is at least as good as that of fixed-site laboratory methods.

Bias estimates were negligible or positive for all beryllium loadings for both PEMs (Table 4), and ranged from 0.0–0.10. In terms of recovery, mean values determined for beryllium loadings for all of the PEM samples (e.g., see Tables 1 and 2) were within ±10 % of the reference values. Typically, recoveries of 100 % ±15 % are regarded as acceptable for meeting the requirements of quantitative analytical methods [20].

A limitation of this study is that this collaborative interlaboratory evaluation did not utilize real aerosol samples generated from beryllium-containing materials. Generally, it is desirable to evaluate methods using performance evaluation samples that are as realistic as possible. But because of the serious health hazards and high costs associated with the generation of beryllium aerosols, it was not deemed feasible to prepare PEMs from beryllium-containing aerosols for this study. It would also be of interest to evaluate the portable fluorescence method on-site in the field, but such an effort is outside the scope of this investigation.

In summary, the results of the interlaboratory evaluation of the field-portable extraction and fluorescence method for beryllium indicate that the method is effective for the quantitative measurement of soluble forms of trace beryllium in MCE and Whatman 541 filter samples. Estimates of within-laboratory and between-laboratory precision compared favorably with interlaboratory precision estimates from a beryllium proficiency testing program, and bias estimates were 10 % or below for each performance evaluation sample tested. Performance data obtained here represent the minimum that is required for NIOSH methods and ASTM International standards. It is intended that future studies will address real-world sample matrices and on-site evaluations of the portable fluorescence method.

Disclaimer

Mention of company names or products does not constitute endorsement by the Centers for Disease Control and Prevention.

Acknowledgments

This work was carried out in coordination with the Analytical Subcommittee of the Beryllium Health and Safety Committee. We thank the following volunteer laboratories for participating in this collaborative study: CDC/NIOSH (Health Effects Laboratory Division), Morgantown, WV; Los Alamos National Laboratory (Isotope and Nuclear Chemistry Group), Los Alamos, NM; Sandia National Laboratory, Albuquerque, NM; and the University of Arizona (Chemistry Department and Materials Science Department), Tucson, AZ. Thanks are also extended to Keith Crouch and Jensen Groff of CDC/NIOSH (Division of Applied Research and Technology) for their kind assistance. We appreciate the valuable comments offered by the referees.

References

[1] Code of Federal Regulations, 10 CFR Part 850, *Chronic Beryllium Disease Prevention Program*, , U.S. Department of Energy, Washington, DC, 1999.

[2] Ruedas Rama, M. J., Medina, A. R., and Díaz, A. M., "Implementation of Flow-Through Multi-Sensors with Bead Injection Spectroscopy – Fluorimetric Renewable Biparameter Sensor for Determination of Beryllium and Aluminum," *Talanta*, Vol. 62, 2004, pp. 879–886.

[3] Wang, J., Tian, B. M., "Trace Measurements of Beryllium by Adsorptive Stripping Voltammetry and Potentiometry," *Analytica Chimica Acta*, Vol. 270, 1992, pp. 137–141.
[4] McCleskey, T. M., Presentation at the American Chemical Society National Conference, Anaheim, CA, Apr. 2004.
[5] Berylliant, Inc., *Manual for Procedures and Kit Description for Determination of Beryllium (BeFinder)*, Berylliant, Inc., Tucson, AZ, Dec. 2004.
[6] Matsumiya, H., Hoshino, H., and Yotsuyanagi, T., "A Novel Fluorescence Reagent, 10-hydroxybenzo[*h*]quinoline-7-sulfonate, for Selective Determination of Beryllium(II) Ion at pg cm^{-3} Levels," *Analyst*, Vol. 126, 2001, pp. 2082–2086.
[7] Pal, B. K. and Baksi, K., "Chromotropic Acid as Fluorogenic Reagent. 1. Fluorometric Determination of Beryllium," *Mikrochimica Acta*, Vol. 108, 1992, pp. 275–283.
[8] Donascimento, D. B. and Schwedt, G., "Off-Line and Online Preconcentration of Trace Levels of Beryllium Using Complexing Agents with Atomic Spectrometric and Fluorometric Detection," *Analytica Chimica Acta*, Vol. 283, 1993, pp. 909–915.
[9] Matsumiya, H. and Hoshino, H., "Selective Determination of Beryllium(II) Ion at Picomole per Deciliter Cubed Levels by Kinetic Differentiation Mode Reversed-Phase High-Performance Liquid Chromatography with Fluorometric Detection Using 2-(2'-Hydroxyphenyl)-10-hydroxybenzo[*h*]quinoline as Precolumn Chelating Reagent," *Analytical Chemistry*, Vol. 75, 2003, pp. 413–419.
[10] National Institute for Occupational Safety and Health, *NIOSH Manual of Analytical Methods*, 4th ed., NIOSH, Cincinnati, OH, 1994.
[11] *Annual Book of ASTM Standards*, Vol. 11.03, ASTM International, West Conshohocken, PA, 2004.
[12] ASTM Standard E 691-99, "Standard Practice for Conducting an Interlaboratory Study to Determine the Precision of a Test Method," *Annual Book of ASTM Standards*, ASTM International, West Conshohocken, PA, 1999.
[13] American Conference of Governmental Industrial Hygienists, *2004 Threshold Limit Values for Chemical Substances and Physical Agents & Biological Exposure Indices*, ACGIH, Cincinnati, OH, 2004, updated annually.
[14] www.sandia.gov/BHSC/subs/analytical.htm (accessed 10 Dec. 2004).
[15] *Form and Style for ASTM Standards*, ASTM International, West Conshohocken, PA 2004.
[16] Ashley, K., Song, R., Esche, C. A., Schlecht, P. C., Baron, P. A., and Wise, T. J., Ultrasonic Extraction and Portable Anodic Stripping Voltammetric Measurement of Lead in Paint, Dust Wipes, Soil and Air – An Interlaboratory Evaluation," *Journal of Environmental Monitoring*, Vol. 1, 1999, pp. 459–464.
[17] ASTM Standard E 1775-01, "Standard Guide for Evaluating Performance of On-Site Extraction and Portable Electrochemical or Spectrophotometric Analysis for Lead," *Annual Book of ASTM Standards*, ASTM International, West Conshohocken, PA, 2001.
[18] Welch, L., Presentation at the Analytical Subcommittee meeting of the Beryllium Health and Safety Committee, Savannah River Site, SC, Feb. 2004.
[19] NIOSH Method 7102, Beryllium and Compounds, as Be; in *NIOSH Manual of Analytical Methods*, 4th ed., NIOSH, Cincinnati, OH, 1994.
[20] Kennedy, E. R., Fischbach, T. J., Song, R., Eller, P. M., and Shulman, S. A., *Guidelines for Air Sampling and Analytical Method Development and Evaluation*, (DHHS [NIOSH] Publ. No. 95–117), NIOSH, Cincinnati, OH, 1995.

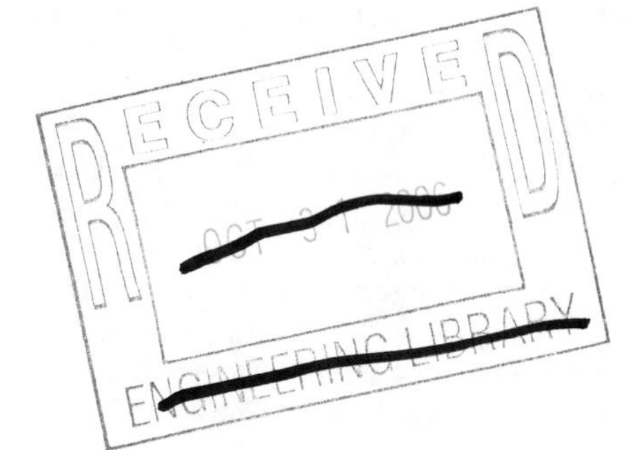
RECEIVED
OCT 3 1 2006
ENGINEERING LIBRARY